T0327357

The Impact of Automatic Control Research on Industrial Innovation

The Impact of Automatic Control Research on Industrial Innovation

Enabling a Sustainable Future

Edited by

Silvia Mastellone
Institute for Electric Power Systems
University of Applied Science Northwestern Switzerland
Windisch, Switzerland

Alex van Delft
VanDelft.IT
Sittard, The Netherlands

Maria Domenica Di Benedetto, *Series Editor*

Published by John Wiley & Sons, Inc., Hoboken, New Jersey.
Published simultaneously in Canada.

For general information on our other products and services or for technical support, please contact our Customer Care Department within the United States at (800) 762-2974, outside the United States at (317) 572-3993 or fax (317) 572-4002.

Wiley also publishes its books in a variety of electronic formats. Some content that appears in print may not be available in electronic formats. For more information about Wiley products, visit our web site at www.wiley.com.

Library of Congress Cataloging-in-Publication Data

Names: Mastellone, Silvia, editor. | Delft, Alex van, editor.
Title: The impact of automatic control research on industrial innovation : enabling a sustainable future / edited by Silvia Mastellone, Alex van Delft.
Description: Hoboken, New Jersey : Wiley, [2024] | Includes index.
Identifiers: LCCN 2023039036 (print) | LCCN 2023039037 (ebook) | ISBN 9781119983613 (hardback) | ISBN 9781119983620 (adobe pdf) | ISBN 9781119983637 (epub)
Subjects: LCSH: Automatic control. | Technological innovations. | Technological innovations–Environmental aspects.
Classification: LCC TJ213 .I5375 2024 (print) | LCC TJ213 (ebook) | DDC 629.8–dc23/eng/20230909
LC record available at https://lccn.loc.gov/2023039036
LC ebook record available at https://lccn.loc.gov/2023039037

Cover Design: Wiley
Cover Image: © Andriy Onufriyenko/Getty Images

Set in 9.5/12.5pt STIXTwoText by Straive, Chennai, India

Silvia Mastellone: To Peter, Julian, and Lorenz, who constantly inspire me to create something beautiful.

Alex van Delft: To Cariene, for her patience, love, support, and understanding why this is important to me.

Contents

About the Editors *xi*
List of Contributors *xiii*
Foreword *xv*
Preface *xvii*
Acknowledgments *xix*

1   **Introduction** *1*
          Silvia Mastellone and Alex van Delft
1.1   Background and Motivation *4*
1.2   The Cradle of Innovation *5*
1.3   Final Remarks *10*
          References *11*

Part I   Infrastructure and Mobility *13*

2   **Data Industry** *15*
          Amritam Das, Ramachandra R. Kolluri, and Iven Mareels
2.1   Introduction *15*
2.2   Anatomy of a Data Center *20*
2.3   Reducing Power in a Data Center *23*
2.4   Codesigning Energy Assets *31*
2.5   Beyond the Data Center *36*
2.6   Conclusion *36*
          References *37*

3   **Building Automation** *43*
          Scott A. Bortoff, Bryan Eisenhower, Veronica Adetola, and Zheng O'Neill
3.1   Introduction *43*

3.2   HVAC Background   *45*

3.3   Industry Trends and Drivers of Innovation   *60*

3.4   Consequences and Implications   *65*

3.5   Industry Needs-Driven Innovation   *67*

3.6   Vision-Driven Innovation   *70*

3.7   Conclusions   *79*
          References   *80*

4   Future Impact and Challenges of Automotive Control   *85*
          Stefano Di Cairano, Carlos Guardiola, Andreas A. Malikopoulos, and
          Jason B. Siegel

4.1   Introduction   *85*

4.2   Internal Combustion Powertrain   *86*

4.3   Electrification   *96*

4.4   Driver Assistance Systems and Automated Driving   *104*

4.5   Connected and Integrated Transportation Systems   *111*
          References   *117*

Part II   Energy and Production   *135*

5   Control of Electric Power Conversion Systems   *137*
          Peter Hokayem, Pieder Joerg, Silvia Mastellone, and Mario Schweizer

5.1   Introduction   *137*

5.2   Power Electronic Conversion Systems   *138*

5.3   Grid-Connected Power Converters   *145*

5.4   Industrial Drives   *157*

5.5   Research Roadmap   *162*

5.6   Conclusions   *164*
          References   *164*

6   Robotics and Manufacturing Automation   *169*
          Alisa Rupenyan and Efe C. Balta

6.1   Introduction   *169*

6.2   Vision   *178*

6.3   Future Challenges and Trends   *188*
          References   *189*

7   Process Industry   *199*
          Alex van Delft

7.1   Introduction   *199*

7.2   Existing Control Challenges in Process Industry   *208*
7.3   Vision of the Future Generation of Products and Processes   *211*
7.4   Formulation of Control Research – Directions for Process
         Industry   *216*
7.5   Conclusion: What Drives and Blocks Innovation in Process
         Industry?   *221*
         References   *222*

         Index   *225*

About the Editors

Silvia Mastellone is a professor of Control and Signal Processing at the University of Applied Science Northwestern Switzerland. She holds a PhD degree in Systems and Entrepreneurial Engineering from the University of Illinois at Urbana-Champaign, a master's degree in Electrical Engineering from the University of New Mexico and a Laurea degree in Computer Engineering from the University of Rome.

She held several R&D positions across different regions at companies including Xerox, Alenia Marconi Systems, and ABB. From 2008 to 2016, she worked as a principal scientist for the ABB Corporate Research Center in Switzerland, where she led research projects and contributed to defining the research strategy in the areas of advanced control and optimization for energy systems.

Her research interests include decentralized control and estimation of networked control systems, applied to sustainable optimal operation and diagnostics for power conversion and energy systems. She served as a member of the IFAC Industry Executive Committee from 2015 to 2023.

She is a principal investigator, an executive member of the NCCR-Automation, and a member of the advisory board for the multiutility company IBB. She currently serves as the VP of Finance for the International Federation of Automatic Control and is a member of the CSS Board of Governors.

Alex van Delft has a master's degree in engineering physics and a PhD in physics/process control and optimization from Eindhoven University of Technology. In 1989, he joined Royal DSM, The Netherlands, and held several positions in Manufacturing, Engineering and (Program) Management. He developed and applied methods to set up an automation strategy for production plants and was engaged in setting up strategic partnerships with suppliers in the field of process automation. In 2005, he moved to DSM's headquarters as Corporate Manager Process Control, a position he held until 2020, where his focus was to strengthen the role of process automation in company-wide operational excellence programs.

In December 2020, he founded VanDelft.IT as an independent consultancy company. The technology portfolio comprises knowledge and support in the areas of process control and automation, machine learning, and data engineering, as well as endeavors to close the gap between theory and practice in process automation.

From 2010 to 2020, he served as the chairman of WIB, the Dutch-Belgian Process Automation End-users Association, working on application standards and practices jointly with European sister organizations in countries such as Germany, the United Kingdom, France, and Italy.

He also fulfilled board roles in measurement and control for organizations including the Royal Dutch Institution of Engineers and the Dutch Association for Post-Academic Education, where he provided training. He has been a member of the IFAC Industry Executive Committee since 2015.

List of Contributors

Veronica Adetola
Pacific Northwest National Laboratory
Richland
WA
USA

Efe C. Balta
Control and Automation Group
inspire AG
Zurich
Switzerland

Scott A. Bortoff
Mitsubishi Electric Research
Laboratories
Cambridge
MA
USA

Stefano Di Cairano
Mitsubishi Electric Research
Laboratories
Cambridge
MA
USA

Amritam Das
Department of Electrical Engineering
Eindhoven University of Technology
Eindhoven
The Netherlands

Bryan Eisenhower
Carrier Corporation
Palm Beach Gardens
FL
USA

Carlos Guardiola
Escuela Técnica Superior de Ingeniería
Industrial
Universitat Politècnica de València
Valencia
Spain

Peter Hokayem
ABB Motion
ABB Switzerland Ltd
Turgi
Switzerland

Pieder Joerg
ABB Motion
ABB Switzerland Ltd
Turgi
Switzerland

Ramachandra R. Kolluri
IBM Consulting Australia
IBM
South Bank
Victoria
Australia

Andreas A. Malikopoulos
Mechanical Engineering
University of Delaware
Newark
DE
USA

Iven Mareels
Institute for Innovation, Science and
Sustainability
Federation University Australia
Berwick
Victoria
Australia

Silvia Mastellone
Institute for Electric Power Systems
University of Applied Science
Northwest Switzerland
Windisch
Switzerland

Zheng O'Neill
Department of Mechanical
Engineering
Texas A&M University
College Station
TX
USA

Alisa Rupenyan
ZHAW Centre for AI
School of Engineering
ZHAW Zurich University for Applied
Sciences
Winterthur
Switzerland

Mario Schweizer
ABB Corporate Research
ABB Switzerland Ltd
Baden-Dättwil
Switzerland

Jason B. Siegel
Mechanical Engineering
University of Michigan
Ann Arbor
MI
USA

Alex van Delft
VanDelft.IT
Sittard
The Netherlands

Foreword

"Control is everywhere," control engineers and scientists are fond of stating, and indeed it is. The complex engineered systems, solutions, and products, whose design, operation, maintenance – indeed, whose lifecycles – rely crucially on control technology, are virtually synonymous with what makes societies, industries, and our civilization run. The list includes aircraft and automobiles, homes and buildings, factories and process plants, biomedical devices and healthcare systems, power generators and transmission grids, and much more.

And yet, or perhaps because of this ubiquity, control is nowhere as well, in the sense that its contribution to the performance, safety, efficiency, reliability, economy, and other critical characteristics of these products and systems is hidden from view. We see a wind turbine rotating, or a rocket launch, or electricity at the outlet, or the internet at the call of a browser, and the first thought that comes to mind is not, "that's control!"

This disconnect between abstract technology and its physical manifestation extends to control technologists as well. On the exploratory end of the research, development, and engineering spectrum, we develop theories and algorithms. On the other end, our output is consumed by a specific application domain or industry subsector. Yet effective products and solutions require collaboration across the scale. Researchers need to have a deeper understanding of how, where, and why control is used. Product developers need to appreciate the relevance of broad-based R&D. Those working in the extended middle must be able to map each end to the other. Students should be apprised of the opportunities open to them, regardless of their theoretical or practical bent. All of us, even beyond the control community, must be able to connect the dots.

The Impact of Automatic Control Research on Industrial Innovation addresses such imperatives. It introduces the "cradle of innovation" as a model for the interplay between research-driven and market-driven innovation, illustrating the dynamics involved with examples from numerous domains. Several industry sectors are surveyed in depth, with analyses of the state of the art, market trends, innovation drivers and obstacles, and technology roadmaps, as well as nontechnological aspects such as culture and ethics.

The seeds for this edited volume were laid in discussions in the IFAC Industry Committee on the topic of innovation in industry applications and the role of control in it. Prof. Silvia Mastellone and Dr. Alex van Delft took on the challenge of delving into this multifaceted topic. This book is the culmination of a years-long effort by the editors, not the least of which is their recruitment of some of the top experts in the world for the industries and sectors covered.

Read the book! Regardless of your area of expertise or theory/practice alignment, if you are interested in control systems and their critical role in ensuring a sustainable future for humanity and the planet, you'll come away with a new appreciation for this ubiquitous and foundational discipline.

Tariq Samad, PhD
Technological Leadership Institute,
University of Minnesota, USA
Founding Chair, IFAC Industry Committee

Preface

This book is the result of a longstanding cooperation between members of the Industry Committee of the International Federation of Automatic Control. It started with the vision and ambition to address the challenge of the academic–industry gap in research and was initially discussed during a workshop at the 2017 IFAC World Congress in Toulouse, France. Subsequently, we conducted an industry-wide survey on the topic of academic–industry cooperation, and we published the results in a Control Engineering Practice article in 2021, in which we introduced a framework for innovation in the field of automatic control. Inspired by the positive feedback received and the request for more industry-specific insights, we decided to work on an in-depth study of different industry sectors. We were fortunate to win the engagement of a board of main authors, both from academia and industry, who committed to work on this edited volume. Meanwhile, we organized webinars and conducted several forum sessions at conferences on the topic, which allowed us to gain more insights into the link between control research, innovation, and technology. We also realized that bridging the gap between fundamental research and application, as well as understanding and enabling an innovation process, are important topics, not only for control engineers and researchers but also for graduate students in automatic control, as they will lead future research and innovation. With this volume, we would like to offer them a vision of the possibilities for their professional paths and impact in academic and industry alike. We realized that there is an important driving force that binds us all: the current societal and environmental challenges demand to employ the best innovative technologies to encompass performance, efficiency, and reliability targets at the service of a sustainable future.

November 2023

Silvia Mastellone
Windisch, Switzerland
Alex van Delft
Sittard, The Netherlands

Acknowledgments

This edited volume would not have been possible without the joint effort and inspirational cooperation between all people, see the List of Contributors, from different industry sectors and academic backgrounds.

We would like to express our special gratitude to Tariq Samad for being the continuous driving force and challenging partner in this process; to the members of the Editorial Board for their great leadership; and to all members of the IFAC Industry Executive Committee: Kevin Brooks, Moncef Chioua, Stefano Di Cairano, Philippe Goupil, Steve Kahne, Iven Mareels, Alisa Rupenyan, and Atanas Serbezov, for the stimulating discussions that contributed to shaping the ideas behind this work.

Editorial Board: Silvia Mastellone (FHNW, University of Applied Science Northwestern Switzerland), Alex van Delft (vanDelft.IT, former DSM, The Netherlands), Tariq Samad (University of Minnesota, USA), Iven Mareels (Federation University Australia, former IBM, Australia), Scott Bortoff (Mitsubishi Electric Research Labs, USA), Stefano Di Cairano (Mitsubishi Electric Research Labs, USA), and Alisa Rupenyan (ETHZ, Switzerland).

1

Introduction

Silvia Mastellone[1] and Alex van Delft[2]

[1]*Institute for Electric Power Systems, University of Applied Science Northwestern Switzerland, Windisch, Switzerland*
[2]*VanDelft.IT, Sittard, The Netherlands*

Technological innovation has shaped human lives across generations, but what are the basic forces driving the innovation process? Arguably we can state that the drive for innovation is rooted in the genuine human curiosity for knowledge, the desire to realize ambitious visions, and, at the same time, in the need for progress and comfort in our daily lives.

Automatic control, as an elegant multidisciplinary science that sets systems in motion, has enabled key steps in the history of technological innovation, from the Kalman filter that empowered humans to reach the moon, to optimal and robust controllers today pervasively present in every system and every process across industry sectors. In an environment where the complexity of engineering systems is ever-growing and technology is developing toward more digital and data-based solutions, automatic control is undergoing a transformation by integrating classical methods with data-driven approaches to address the new complexity, thus opening the door to a new chapter in its history. In this context, it is valuable to identify the way automatic control can enable the next innovation steps in different industrial sectors and thus realize its full potential. To address this question from an application perspective, in [1] we proposed a framework at the interplay between incremental improvement and disruptive innovation. The framework, named *the cradle of innovation*, will be presented in Section 1.2 and consists of a sustainable innovation process driven by a long-term vision and market requirements, where system know-how, economical and technical requirements are considered to ultimately bring a brilliant idea into practice.

The work presented in this volume is part of a broader ongoing effort within the IFAC Industry Committee formed by academic and industrial members and established by IFAC in 2017 with the objective of bridging the gap between industry and academia in the field of automatic control.

Besides providing a framework for the innovation process, the scope of the paper [1] was to link automatic control research to technology innovation. Within this scope, different industrial sectors and government institutions were surveyed, the data were analyzed and translated into technical requirement specifications. Finally, the paper provided pointers to research directions that would address the sustainability challenges across industries.

Starting from this point, with the present volume, we aim to apply the framework of *the cradle of innovation*, expand and detail this concept across six industry sectors.

Building on this vision, in the present volume we invite the reader to join a journey toward the birth of innovation across six specific industry sectors. The journey is inspired by a story that took place in the eighteenth century; the story of the Turk [2], an eighteenth-century automaton that could beat human chess opponents (see Figure 1.1).

The Turk first appeared in Vienna in 1770 as a chess-playing robot dressed in Turkish clothing, seated above a cabinet with a chessboard on top. The operator would assemble a paying audience and invite a challenger to play chess. The automaton would gaze at the opponent's move, ponder, then raise its mechanical arm, and make a move. Of course, the thing was a hack – a clever magician's illusion. The only real ingenuity was a hidden chess player inside the machine.

It is true that the late eighteenth century was a great age of automatons, but the deeper truth that chess-playing was an entirely different kind of creative activity seemed as obscure to people at that time as it seems obvious to us now.

The great-grandfather of computer science, Charles Babbage, saw the Turk and though he realized that it was probably a magic trick, he also asked himself what exactly would be required to produce an elegant solution. What kind of technology would one need to develop in order to build a machine that plays chess? And his "difference engine" – the first computer – rose in part from his desire to believe that there was a beautiful solution to the problem, even if the one before him was not.

Taking inspiration from the story of the Turk, with this volume, we ask the same question for the next generation of products, processes, and services across several industrial sectors: What does the future look like? What is beyond hacking? What would an elegant solution look like?

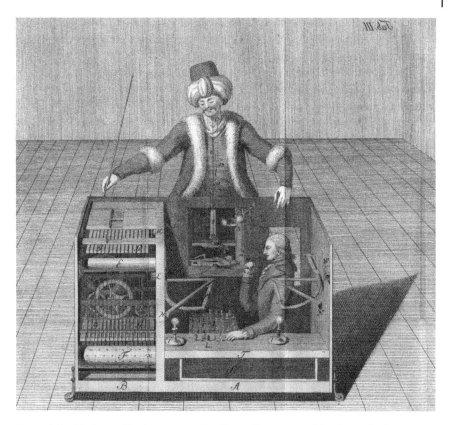

Figure 1.1 Mechanical Turk or Automaton Chess Player was a fake chess-playing machine constructed in the late eighteenth century. Source: Joseph Racknitz/Humboldt University Library.

The volume includes six chapters and is organized into two main parts: Part I focuses on Infrastructure and Mobility and includes the following:

- Data Industry
- Building Automation
- Automotive Control

Part II addresses Energy and Production and includes:

- Power Conversion Systems
- Robotics and Manufacturing Automation
- Process Industry

Each chapter will discuss drivers and limits to innovation for a specific sector. Starting from customer needs and challenges, and system requirements, an applied research agenda will be formulated.

In addition to the research directions driven by industrial requirements, there are visionary ideas that promise to spark a new drive for innovation and where automatic control plays a pivotal role. Examples of such disruptive visions include the *city of the future* characterized by pervasive automation in the *transportation* (e.g. hyperloop and autonomous cars), *energy* (e.g. autonomous microgrids and H_2 economy), *manufacturing* (e.g. Industry 4.0), and *financial sectors*. Additionally, the adoption of control concepts in support of management decision-making could open completely new dimensions with great benefits for both fields.

1.1 Background and Motivation

The gap between fundamental control research and practice has been addressed by several authors from different perspectives. In 1964, Axelby [3] observed that "Certainly some gap between theory and application should be maintained, for without it there would be no progress.... It appears that the problem of the gap is a control problem in itself; it must be properly identified and optimized through proper action."

In a paper by Bennett [4], a historic overview is given of the landmark developments in automatic control. It began in the nineteenth century, when developments were mainly driven by industrial problems, e.g. the steam engine governor. Later on, the PID controller was developed by Elmer Sperry. The first theoretical analysis of a PID controller was published by Nicolas Minorsky in 1922. Another development highlighted in the paper is the feedback amplifier that enabled long-distance telephony, combining experimental data and mathematical models. In the era of classical control theory, the focus was on the development of rigorous mathematical foundations. Later on, the development was driven and sponsored by aerospace and defense, and the advancements in computing power allowed to solve more complex problems.

Rosenbrock, in his work [5], addresses the dilemma of whether automatic control should further develop toward fundamental theory backed up by rigorous mathematics or engineering more centered around experience and intuition. He points toward future developments where computers enhance the human skills rather than replace them.

Aström and Kumar [6] describe the dynamic gap between theory and practice as rooted in the open-loop process of theoretical research without feedback from practice. With current technology, deployment and implementation of complex control solutions have become simpler, thus reducing the gap between theory and application.

Lamnabhi-Lagarrigue et al. [7] build on this analysis and bring it a step further by describing the cross-fertilization and bi-directional interplay between five critical societal challenges (transportation, energy, water, healthcare, and manufacturing) and seven research and innovation challenges (cyber-physical systems of systems, distributed networked control systems, autonomy, cognition and control, data-driven dynamic modeling and control, cyber-physical and human systems, complexity and control in networks, and critical infrastructure systems). The main recommendation from their analysis is the fostering of both fundamental and application-oriented research in sector-specific programs and in ICT as a program that provides enabling technologies for all sectors.

In the paper by Deng [8], the author provides an overview on developments and application areas in automatic control that are driven by societal challenges such as food production, land use, water, logistics, and e-health.

In his 2020 editorial, Grimble [9] establishes a concise link between historical developments in automatic control and the need for a broader, systems-engineering-driven approach.

In summary, the evolution of automatic control has been driven so far by industry, the requirements for theoretically rigorous foundations, aerospace, defense, and the need to address various societal challenges.

In this volume, we aim to further establish control as a discipline that enables innovation in technology by analyzing the innovation dynamics in more detail for specific industry sectors. We introduce a cyclic process for innovation based on [1], where ideas evolve through various stages of selection and transformation and are finally brought to life. Within this process, we identify barriers, enablers, and key drivers for the process in various industry sectors, then through a thorough analysis, those drivers are linked to system requirement specifications and finally to a control research framework or roadmap.

1.2 The Cradle of Innovation

1.2.1 A Framework for Innovation

To establish an innovation enabling framework, it is required to identify factors that affect innovation. To this end, we consider two innovation processes depicted in Figure 1.2. The first process, referred to as *research-driven innovation*, starts from an abstract idea, a theoretical concept, that is transformed and finally realized in an application (product, process, or service). The second process, referred to as *market-driven innovation*, starts from customer requirements that define concrete required technology developments and leading to a research portfolio.

In the first process, *"from research to realized application,"* a preliminary idea is proposed without considering technical feasibility and financial benefits. The idea is then developed and matured through different stages to be finally implemented in a product or process. At each stage, the idea undergoes a transformation and often does not survive the feasibility and profitability tests that are posed at each stage.

In the process *"from customer needs to research focus,"* the starting point is the customer intended as the end user of a specific technology, the market, and in a broader sense, society and its needs. The customer might not have know-how about the technology, but he or she can provide user requirement specifications for a product or process, that is, what are concrete characteristics that he or she would like to see in the product. Those specifications are then translated into product requirement specifications and finally into technical requirement specifications.

In both approaches, once a vision of the next generation of product, processes, or services is formed, the next step is the identification of the key challenges toward the realization of the vision.

The flow in the cyclic innovation process described in Figure 1.2 is catalyzed by systematically translating customer requirements into technical requirements and finally populating the research portfolio. Similarly, an idea is matured through a multi-stage transformation process, where profitability and feasibility criteria are considered while shaping the idea from one stage to the next, until its realization into practice. This requires properly balancing the research agenda so as to include fundamental and implementation aspects.

Vision-driven innovation tools, like design thinking, but also agile and scrum methods, serve to increase the effectiveness and speed of the idea transformation process at each stage.

Figure 1.2 From research to realized application, from customer needs to research focus.

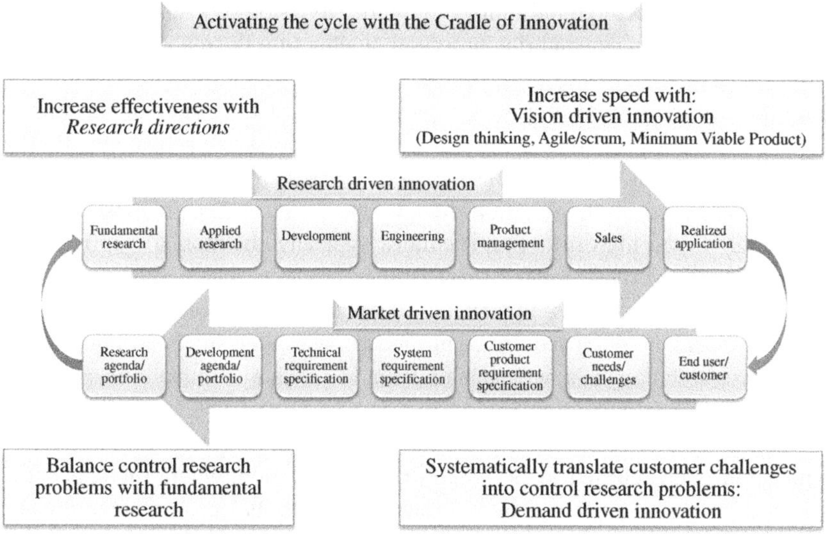

Figure 1.3 The framework to close the gap, enabled by the cradle of innovation.

In both processes, we can additionally characterize innovation as disruptive or incremental. Disruptive innovation is mostly guided by a long-term vision that looks beyond the existing technology, and it is typically accompanied by larger risks. Examples of such disruptive innovations are the touch screen (invented first by IBM but really made disruptive by Apple's Steve Jobs) and the Solar-X program.

Incremental innovation is characterized by smaller improvements in the current technology as it takes into account the constraints and limitations in implementing the innovation, and it is a structured process and requires analysis of each step. It is, however, limited in its possibility to accommodate substantial innovation.

In the case of incremental innovation, the probability of successfully driving an idea into the market is estimated to be up to 60–75% for an innovation using existing technology in the company and intended for the company's current market, see [10]. This success rate decreases significantly to 5–25% for "out of the box" innovation.

Disruptive innovation is such a rare stone and without proper grounding in the majority of cases, the initial idea dies at some point between the vision and the implementation phase. On the other hand, the incremental innovation without a long-term vision can bring a technology to complete alienation as non-properly planned incremental steps will accumulate creating an unmanageable complexity.

Combining an incremental innovation with the vision of a long-term solution can lead to a sustainable and rich process that allows for the realization of a minimum viable product that can accommodate subsequent innovation steps. Starting from the two innovation processes depicted in Figure 1.2, the cradle of innovation offers the means to link the two in a circular process and activate the flow as depicted in Figure 1.3.

Disruptive innovation or vision-driven innovation is rare in most industry sectors due to the high risks that it entails. Typically, the most disruptive innovative sectors are those related to consumer products where there is enough demand for novelty and less for reliability. The trade-off between innovation and reliability seems to often require compromises, one interesting exception in the automotive sector is Tesla, where high demand drives disruptive innovation but also addresses safety requirements.

The literature on innovation processes is widely dominated by vision-driven innovation often referred to as design-driven innovation, where the concept of design thinking, as explained by T. Brown [11], with its focus on creativity and experimentation, plays a fundamental role; see [5]. Often those approaches to innovation begin with a brainstorming phase based on the dream question: imagining to wake up five years from now and all of industry and societal problems have been solved; how does this vision of the future look like? Some examples of those visionary ideas are: man to the moon, iPhone, touch screen, bullet trains, and flying reconfigurable cars running on solar energy. Realizing such visions will require an extensive combined effort from several interdisciplinary fields, from fundamental to applied results.

1.2.2 What Drives Innovation?

The probability to successfully introduce a new technology in the market is correlated to the measure in which the technology meets customer requirements at affordable time and cost. This principle is reflected in a standard product development process, where the customers are surveyed about the limitations of the current product and the desired features for the next generation. Based on those inputs, product requirement specifications are defined. In the second stage, those requirements will be translated into technical system requirement specifications by asking the critical question: what would it take to make it happen? Here a combination of creativity and technical know-how is required to understand possibilities and limitations.

Some key drivers for the next-generation technology that have been identified across industry sectors are: cost, time to market, energy, efficiency, process availability, performance, quality, reduction of variability, throughput, yield, sustainability footprint, and digitalization.

Different sectors exhibit specific innovation drivers related to the nature of their business, some examples of key differentiating factors are: B2C versus B2B business, market and business size, competitive versus niche markets and businesses, with or without safety requirements. Those factors determine to a large extent the dominance of one or more drivers. Interesting differences across sectors are the focus on quality and reliability, for example, in the aerospace sector, where factors such as safety and human psychology play a dominant role. In sectors where safety does not play a dominant role cost and time to market tend to be key drivers. This is typically characterized by sectors that focus on consumer products but not exclusively.

Other interesting differences can be observed in robotics, with the main focus on productivity, and IT, with a focus on time to market; as typical consumer product businesses, the high competitiveness requires agile development. In the energy, oil and gas sector, cost, and reliability play a dominant role in addition to availability. In some applications, the optimality of the process performance is secondary with respect to the process availability. As an example, for a power converter driving a gas pipe, every hour of inactivity leads to major losses or blackouts in an electric grid. For the process industry, cost and quality dominate the scene, here, the proximity to consumer product defines the high priority of quality. The drivers presented here provide a lighthouse to identify the direction of the research effort; the next step is to determine the path to reach this goal and specifically identify what are the obstacles in the way.

1.2.3 Challenges Toward Enabling Innovation

Identifying innovation drivers contributes to shaping a vision and defining a direction for technological innovation. The next step is the identification of the obstacles toward the realization of the vision. From the survey results reported in [1], the following limiting factors related to technology have been identified across different industrial sectors: *abundance of data – but limited contextualization, data acquisition from the field and data reliability, design and development time, agile approach, complexity of system and solution, solution integration within the full process or product, security*, and *cost*. Additional context-based points have to be considered that are not directly related to technology, but represent obstacles toward establishing the innovation processes. Some examples are: maturity of the industry and its adaptation to the deployment of new technology, training of developers and operators, legacy processes, change management, open platforms across vendors, IT, human factors, and market acceptance.

Similarly, we can identify innovation enablers that are beyond technology and related to societal factors. Starting from the education system, we may ask whether we are shaping the new generation to be free thinkers and innovators and whether we are offering stimulating study and work environments. To innovate requires thinking out of the box, exploring nontrivial directions as well as a comprehensive system understanding and knowledge of the process through which an idea is implemented in a product.

Business and industry broadcast that future-ready employees need to have multiple areas of expertise or at least appreciate how a range of skills fit together. Grimble [9] especially highlights the need for control engineers to have additional skill sets, including broader system understanding, implementation aspects, application knowledge, and economic aspects, to identify potential and limitation.

Additionally, a greater need for the education system has been recognized in order to integrate science, technology, engineering, and maths (STEM) concepts with the arts (STEAM) across the wider curriculum. Control design is also an "art" [5]. Human minds excel in pattern recognition, assessment of complicated situations and have an intuitive leap toward new solutions. Those skills should be cultivated in young innovators. As for the work environment, as argued in the Free innovation paradigm [12], companies like Google have been experimenting with ideal environments for creation, with large spaces for thinking, discussing, and generating ideas. But there is more when it goes to motivation and creation. A series of studies on work motivation carried out at MIT, and summarized in [13], describes the intrinsic nature of human motivation, highlighting the main aspects that drive sustainable motivation: autonomy, mastery, and purpose. The author argues against old models of motivation driven by rewards and fear of punishment, dominated by extrinsic factors such as monetary reward. Finally, the drive for innovation does not stop at the formulation of an idea, the knowledge, and capability to bring the idea into the real world requires the alignment of economic and technical requirements. This process can be simplified if the idea was originally conceived with the techno-economical aspects of the end product.

1.3 Final Remarks

With this volume, we offer an industrial perspective on the future of control research, highlighting its impact on technological innovation and opportunity for technology transfer. The main scope is to create a bridge between the control research community and the various industry sectors.

The volume is dedicated to three main groups: (i) the scientific and technical control community, including researchers and control engineers in academic, government, and industrial institutions. For this group, the volume offers a possible research agenda leading to sustainable technological innovation. (ii) Industry representatives: product managers, project managers, and business owners who are aware of the key innovation steps required in their specific fields. This group has a vision for the future product/process/service and wants to learn how it can be enabled. (iii) Academics that use the volume as reference material for graduate courses or continuing education, e.g. graduate course: "Control practice and its impact in the future of industry." The volume provides students with links between theory and practice and insights into the various industry sectors where control can enable technological innovation.

Finally, the next chapter in the history of technological advancement has to consider the reality of limited natural resources. A large portion of the industry will focus in the next 10–20 years mainly on moving from fossil fuels to electricity (energy transition) and further reducing the ecological footprint, according to the UN's sustainable development goals. But energy is only one of several limited resources we rely on, water, mineral, energy, and biological resources will pose our next challenge.

Performance and efficiency can no longer be the only criteria considered for innovation. Sustainability has to become part of our objectives, constraints, incentives, and decision making when we engineer new solutions.

Automatic control, as a rigorous discipline that connects the foundation of elegant mathematics with the application aspects of engineering, has a pivotal role in orchestrating the multidisciplinary group to address the societal and technological challenges for a sustainable future.

References

1 Silvia Mastellone and Alex van Delft. The impact of control research on industrial innovation: What would it take to make it happen? *Control Engineering Practice*, 111:104737, 2021.

2 T. Standage. *The Turk: The Life and Times of the Famous Eighteenth-Century Chess-Playing Machine*. Walker and Company, New York, 2002.

3 G. S. Axelby. The gap—form and future. *IEEE Transactions on Automatic Control*, 9(2):125–126, 1964.

4 S. Bennett. History of automatic control to 1960: An overview. *IFAC World Congress*, 1996.

5 H. H. Rosenbrock. The future of control. *Automatica*, 13(4):389–392, 1977.

6 K. J. Aström and P. R. Kumar. Control: A perspective, survey paper. *Automatica*, 50:3–43, 2014.

7 F. Lamnabhi-Lagarrigue, A. Annaswamy, S. Engell, A. Isaksson, P. Khargonekar, R. M. Murray, H. Nijmeijer, T. Samad, D. Tilbury, and P. Van den Hof. Systems & control for the future of humanity, research agenda: Current and future roles, impact and grand challenges. *Annual Reviews in Control*, 43:1–64, 2017.

8 W. Deng. Future control and automation. *Proceedings of the 2nd International Conference on Future Control and Automation*, 2012.

9 M. J. Grimble. Welcome to 2020 and the future of control. *Advanced Control for Applications: Engineering and Industrial Systems*, Volume 2, 2020.

10 G. Day. Is it real? Can we win? Is it worth doing?: Managing risk and reward in an innovation portfolio. *Harvard Business Review*, 2007.

11 T. Brown. Design thinking. *Harvard Business Review*, June 1–9, 2008.

12 E. von Hippel. *Free Innovation*. MIT Press, 2017.

13 Daniel H. Pink. *Drive: The Surprising Truth About What Motivates Us*. Riverhead Hardcover, 2009.

Part I

Infrastructure and Mobility

2

Data Industry

Amritam Das[1], Ramachandra R. Kolluri[2], and Iven Mareels[3]

[1]*Department of Electrical Engineering, Eindhoven University of Technology, Eindhoven, The Netherlands*
[2]*IBM Consulting Australia, IBM, South Bank, Victoria, Australia*
[3]*Institute for Innovation, Science and Sustainability, Federation University Australia, Berwick, Victoria, Australia*

2.1 Introduction

This chapter deals with the emerging platform industries that support and enable the digitization revolution,[1] with a particular attention to address its contribution to enabling a sustainable future, without losing sight of its own significant ecological footprint.

Digitization is at the essence of the fourth industrial revolution, which is also often referred to as the era of cyberphysical systems and discussed in terms of Society 5.0 and Industry 4.0. This revolution is fueled by technological advances in Artificial Intelligence (AI) and process automation more generally (all other chapters in the book underscore this). It evolves from and builds on the third industrial revolution, which is punctuated by the evolution of digital electronics, and the control as well as automation of manufacturing. Indeed the present potential to collect, communicate, and analyze data appears almost limitless. Seen optimistically, a well-implemented cyberphysical world promises to drive decisions and actions based on factual evidence (i.e. data from the past) astutely combined with modeling (as in computer models to predict plausible futures), which is nowadays typically supported by AI and computer aided design tools. This optimism should be tempered by the realization that cyber-crime equally benefits from these exponential opportunities, and that robotic, drone armies are

---

1 In the rest of this chapter, when reference is made to data, it is understood to refer to digital data.

The Impact of Automatic Control Research on Industrial Innovation: Enabling a Sustainable Future,
First Edition. Edited by Silvia Mastellone and Alex van Delft.
© 2024 The Institute of Electrical and Electronics Engineers, Inc. Published 2024 by John Wiley & Sons, Inc.

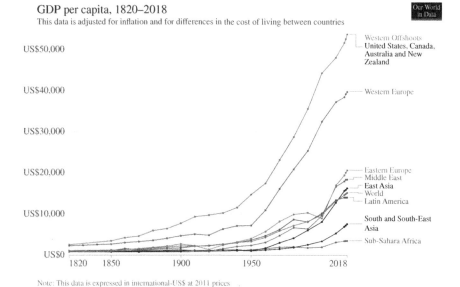

GDP per capita, 1820–2018

This data is adjusted for inflation and for differences in the cost of living between countries

Figure 2.1 Gross domestic product (GDP) per capita data as per the Maddison Project. Source: Data taken from [1].

presently deployed in armed conflict zones, and not a mere nightmare from a dystopian future.

Equally sobering is the well-documented realization that the fourth industrial revolution continues to grow the gap between the developed and the developing world [1], as illustrated in Figure 2.1, and more broadly between those individuals, or corporations that have access to these technologies and those that lag in this access. Particularly, pertaining to the data industry, any systemic asymmetry in the capacity to analyze data (i.e. to extract relevant knowledge or insights from data) leads to significant economic differences, and pronounced societal inequalities [2]. In the context of digital, exponential technologies, the same idea is captured by IBM's current CEO Arvind Krishna as *Winner takes all*. Given the inherent speed at which digital technologies develop, these inequalities become profound, perhaps irreversibly so, very rapidly.

In this context, education and appropriate policy settings have an important role to play. Symbiotically, AI carries with it the promise to enable the formulation of better fact-based policy-making. Equally, digitization's underlying capacity to scale can introduce a new era of greater productivity in the teaching and learning sector allowing us to realize learning outcomes at a scale not realized to date [3]. The ethical and responsible development of our next technologies is equally key here.

The scale of the fourth industrial revolution is perhaps best understood from the investments that are made to enable digitization. These investments are second only to the investments flowing into the decarbonization of the energy and mobility sector. McKinsey's Technology Trends Outlook 2022, see also McKinsey's Quarterly March 2023, estimates annual investments into digitization to be of the order of US$ 600 billion annually, with notable present day expansion in this investment in the world of AI and Cloud/Edge computing services. (The Stanford AI Index Report 2023, estimated the private, corporate, investment into AI development just under US$ 100 billion in 2023.) In comparison, the McKinsey Technology Trends Outlook 2022 report estimates the investments in energy and mobility to be in the order of US$ 700 billion. In our current economic climate, these investments can only be justified on the basis of significant productivity gains (or perceived productivity gains). The prevailing mantra is indeed that the data combined with the AI revolution (that accelerates the scientific modeling, and process automation efforts) affects all economic activities, and provides significant productivity gains. It is said that any economic activity that is not shifting over to these so-called exponential technologies, is facing oblivion.

The economic success of AI is discussed in [4] and identified succinctly as due to the significant reduction of the cost of modeling and in particular models for prediction (under uncertainty), with a concomitant increase in the value of the data used to produce the models.

The recent interest in ChatGPT (a generative pre-trained transformer model to process natural language queries and that respond in a natural language manner[2]) underscores this observation of productivity gain, and economic success, as in less than two months the number of users of ChatGPT grew from a team of developers to over 100 million. The OpenAI group,[3] creator of ChatGPT and similar models envisions ChatGPT as a general-purpose technology that has the potential to significantly affect how we work, see [5]. This ChatGPT activity further points to the need for talent management as digital technologies bring automation to so many aspects of human work.

Digitization with AI appears to parallel the transformation automation and control engineering has brought to all aspects of manufacturing and process industries. This parallel may suggest that the productivity gains are not only realistic, but that there will be sufficient wealth creation to benefit all, and to overcome the initial effects of job displacement. The parallel is not complete, however, as the digitization revolution affects all jobs, including scientific

---

2 ChatGPT is not only well-versed in natural languages, but it is equally credible in producing computer code, increasing the productivity of software development. It can also respond to image prompts.
3 https://openai.com/research/gpts-are-gpts.

discovery, and design jobs, but arguably affects more the classic, white collar, jobs (see also [5]) where (relatively) simple and repetitive cognitive tasks, and associated administrative processes are readily automated. It follows that humans will be required to exercise more judgment, use intuition more often and focus more on synthesis across multiple domains of knowledge. These areas, intuitions, judgment, and synthesis remain elusive for our present AI technologies and available computing resources. Nevertheless, as these technologies progress, and the future of computing includes quantum computing, some of these present barriers will be challenged. There are though fundamental limits to what AI can achieve. These limits suggest that even the most advanced AI technology based on mere correlation analysis of data may never reach the realms of general fluid intelligence in a reliable, and trustworthy manner, see for example [6]. This does not take away from the fact that present-day AI can already be used by experts to push the boundaries of science with significant advantage and allow us to accelerate the scientific discovery process. The productivity gain that can be achieved by judicious use of AI is already impressive across a very large range of human activities. This fundamental shift in the meaning of human work presents significant challenges, as most of us derive a lot of meaning from and through the work we perform.

Data, digitization, and AI trends are staggering, as illustrated by the increasing energy consumption in data centers across the globe [7]. As indicated in [7][4] there is much work that has been done to reduce the power and energy footprint of data centers, but there is more that can and should be done. The broader information technology (IT) sector, including the compute and communicate aspects of data use, and the electronics manufacturing, is presently responsible for about one-tenth of global energy use (roughly equal to half the energy share associated with the transport sector), about 40% of this can be attributed to data and compute centers, another 40% is due to manufacturing of electronics. Moreover, given the large investments in data infrastructure, the total energy requirements for the digital revolution are growing rapidly. Presently, the largest data centers have power ratings in the order of 100 MW. In [8] the authors estimate that the power and energy consumption in large data centers in the USA is rising at around 10% per annum, a trend predicted to last through 2030. The associated annual investment in new data centers is estimated to exceed US\$ 50 billion per annum by 2030, which is only a fifth of the investment in edge solutions, where the majority of the data will be collected and analyzed. As the data volume is growing rapidly, and more opportunities arise for digitization, the actual growth

---

4 A more recent discussion can be found here https://www.numenta.com/blog/2022/05/24/ai-is-harming-our-planet.

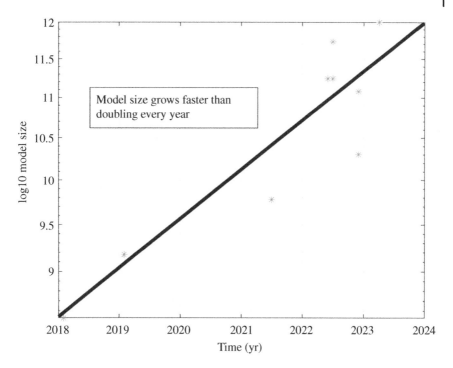

Figure 2.2 Number of free parameters in language models over the last five years.

of the energy required to support the fourth industrial revolution can only be underestimated. This underscores the need to carefully consider how we manage the power and energy requirements for digitization. There is a lot of scope to contain the power and energy use through good control of the available resources.

The use of AI, and in particular the present focus on ever larger foundational models (AI models are already using more than 100 billion tunable parameters, and some are aspiring to use more than a trillion parameters) underscores that the above trends may even be conservative. The larger the data sets and models become (model and data requirements double every 10 months [9], see also Figure 2.2), the more computation their training requires (roughly scaling quadratically with size), and the more asymmetry there exists between those that can develop these frontier AI technologies and those that cannot. Large corporations are now well ahead of even the largest academic and public research institutions in developing new AI, and large models.

In [9], current generative models are identified as consuming in the order of 1 GWh of energy per training round. To put this in context, estimating the power consumption of the human brain at around 10 W, the training of a single generative AI model like ChatGPT equates to more than 10,000 years of continuous (human) brain work.

It is the hypothesis that control and automation, which in some sense form part of this digitization sector, can make a real difference to enable a more sustainable future for the world and digitization specifically. This optimism is shared by many in the industry: *IT electricity usage will rise by 50 percent by 2030. Yet this doesn't have to lead to more carbon dioxide emissions. With huge potential to decarbonize data centers, transport, and buildings – digital is part of the solution (Philippe Delorme, Executive Vice-President, Energy Management, Schneider Electric)*. Nevertheless, as the research in [7] shows, much of this promise is yet to be realized. This was also the thesis put forward by [10], where the authors rather disappointingly conclude that *There is a huge gap between most operators' high levels of energy consumption and costs and the lower levels that are technically possible, even today*. In part, this is due to the data/cloud business model, where the normal operational costs can be passed onto the consumer of the data and cloud services, whereas the infrastructure premium associated with a green data center, and/or highly efficient data center are much more difficult to monetize.

In this chapter, we consider some of the opportunities for automation and control to realize a more sustainable digitized economic future, by considering only one aspect of the digital revolution, namely the data center. We consider the energy and power footprint from both the infrastructure as well as the operations point of view. Minimizing the power and energy footprint is essential as the world transitions to renewable energy. This objective does not change, even when all electrical power is derived from renewable sources. Indeed a circular economy point of view indicates that we need to minimize the overall infrastructure, both from a materials perspective as well as operational power and energy requirements. The aim is to reduce,[5] through design and operational principles, the energy and material footprint that is required over the life-cycle of the data center's assets, from manufacturing, maintenance, and operations to recycling and waste management. The re-conception of the data platform industry from a circular economy, minimal ecological footprint point of view, is a truly complex task, where AI, modeling complex systems, and optimization technologies, can be leveraged to advantage to chart a more sustainable future for the industry.

2.2 Anatomy of a Data Center

Data centers are complex IT environments that provide compute, data storage, and data migration as services (IT workloads) on demand that are flexibly consumed

---

5 Best, or optimal is a contentious notion in this context, the pareto optimal set for minimal energy and material footprint for a data service is not known.

as required. For an overview, with an emphasis on energy consumption, see for example [11].

At the core are the devices to control (store, migrate, encrypt) data and execute compute workload requirements, these are large compute and data servers, supported by uninterrupted power supplies, and a large variety of ancillary services, and network facilities. These are invariably housed in appropriately climate-conditioned environments, with strictly regulated access for both people as well as data (and code) flows.

The raison d'etre for data centers is that data do represent the key assets for most corporations and individuals alike. Hence it pays to entrust specialist providers to protect and maintain the data, as well as the insights that can be derived from them. Such arrangements typically facilitate better resilience, privacy, and security than could be otherwise achieved. In general such third party arrangements provide peace of mind through a professionally managed risk profile, that is to be preferred over self managed data and compute facilities, which is the realm of but the largest organizations.

In terms of energy consumption and data center facility efficiency, the de facto standard is to talk about the Power Usage Effectiveness (PUE), which is the ratio of the *entire* energy use of the data center facility over the energy use of the IT workload, measured over a given period of time. Present best-in-class data centers [9] achieve a PUE of 1.1 measured over a year.[6] Given that all power used in computing essentially degrades to mere heat (low-quality or low-temperature heat), and electronics have to be kept (well) below 308 K (35 °C or 95 °F) and in an environment around 50% humidity to guarantee an acceptable life-time of the components, cooling is key. It follows that a low environmental temperature (and low humidity) is really advantageous, and hence the consideration to float data centers in the ocean, or anchor them on the sea floor, or locate them near the arctic circle. Resilience and disaster recovery plans often require that data centers and compute workload have to be split over different geographic zones, in which case, one also has to take the data migration energy into account (but this is typically much less than the compute energy). In the present hybrid world, the compute infrastructure for a corporation, even an individual will be split over many different data centers, and be constituted as a complex mixture of rented, owned, and infrastructure as service elements. Although the financial and contractual arrangements embedded in these arrangements are important, we focus on the data center as the key infrastructure element in the data platform industry, and how one may lower its power and energy footprint.

In the long term, from a sustainability point of view, it is not the energy, but the carbon footprint associated with the data center (over its entire lifetime) that

---

6 Seasonal, daily variations are to be expected as the environmental temperature and humidity affects cooling efficiency.

matters. First, the embedded carbon footprint is substantial, as electronics carry a significant carbon footprint and the data center's electronic facilities are substantial. There is also a carbon footprint associated with cooling. Given that all these facilities have a finite life-time, it pays to consider the circular economy aspects of the materials and physical assets, servers, batteries, cables, building materials, and so on. This is an area that deserves much more attention. There are fundamental limits (think of the second law of thermodynamics) to how well recycling can be achieved and how much energy is required in manufacturing, maintenance, and recycling. These processes, and design limits need to be understood better to be able to come to a good understanding of what is truly the most sustainable data center in a circular economy sense. Systems thinking is key in this context. Moreover, in this context codesign, that is sizing an asset taking into consideration how it is operated (controlled or managed), maintained, repaired over its lifetime, and eventually recycled, plays a key role. The codesign topic is explored in some more depth in the below, but not the complex cradle-to-grave problem introduced here. In its totality, the design of the data center from a circular economy point of view remains fertile territory for much further research and development.

The carbon footprint associated with power and energy consumption in the data center is much easier to model than the embedded carbon footprint. It is this *easy* carbon footprint that is more often discussed. To eliminate the energy carbon footprint, it suffices that the data center be supplied from purely green or renewable energy sources. The idea of a self-contained microgrid supplied from solar with appropriate storage to support a large data center may not be feasible.[7] It also demands a significant investment premium over traditional grid-connected data centers, and does not come with the same level of power security (although this can be compensated for through IT and data redundancy measures) and potentially a significantly increased embedded carbon footprint. For grid-connected data centers, many data center providers or their consumers purchase carbon offsets to achieve a carbon-neutral IT environment (often mainly for the IT workload, sometimes also including the embedded carbon footprint). No matter what approach is taken to achieve carbon neutrality, it pays to achieve an outstanding PUE and to minimize the actual power requirements for the IT workload. Indeed, the larger the IT workload, and the larger the PUE, the larger infrastructure is required to supply both (peak, average) power and (total) energy. From a circular economy and sustainability point of view, where we look at the energy and material consumption over the life-time of the services and the facilities it matters therefore to minimize the power and energy requirements

---

7 At 100 MW average power, the data center proper would cover some 3 Ha, but the solar array would need to peak at 700 MW and cover an area of 3000 Ha in sunny Western Australia, and require a 1 GWh battery!

for executing the IT workload, although this will not be sufficient to realize a maximally sustainable data center. This follows from the simple observations that circular economy considerations must and can consider more degrees of freedom than simply energy or power minimization.

2.3 Reducing Power in a Data Center

Chapter 3 deals with Building Automation, its observations apply equally to data centers, after all data centers are buildings, and cooling is key in all data centers, and hence smart cooling and building management matters. Given that best-in-class data centers achieve a PUE of 1.1 data centers are a prime example of what can be done with smart building automation.

In this chapter, we therefore consider what can be done with the actual IT workload, which in fact is responsible for about 90% ($1/PUE$) of the power and energy load. Moreover, as we lower the power and energy requirements for the IT workload, we reduce the overall power requirements, or allow for an increased IT capacity for a giving cooling capacity.

Power and energy can be considered at various time and spatial scales associated with the IT workload:

1. At the micro-level power can be orchestrated at the chipset level at near clock frequency in response to the type of workload, and its priority.
2. At the intermediate space/time scale, distributing IT workload over servers, racks, and alternative data centers in response to carbon signals from the grid, accounting for the service level requirements associated with the IT workload, presents opportunities at the time scale of minutes to days, from a rack to a cluster of data centers under management. An example of such control is presented in the paper [12] where the authors propose a thermal-aware scheduling algorithm for batch jobs in geographically distributed data centers. The goal of the proposed algorithm is to reduce energy consumption and thermal hotspots in the data centers while meeting job completion time and network bandwidth constraints. Similarly, at the server or rack level, managing the sleep state of servers provides significant power-saving opportunities whenever IT load is predictable over the time horizon required to wake the server up.
3. Finally, at the scale of developing new data centers, significant gains can be achieved by considering the codesign of the power and cooling services using (dynamic and operational) models for IT workload power requirements over time. The hierarchy of these time/spatial scales allows us to consider the various options to maximize the power savings at each scale separately.

2.3.1 Chip and Chip Management

The evolution of chipsets is characterized by ever-decreasing electronic feature dimensions. Dennard's scaling, a derivative of Moore's law, captured that performance (in terms of GFlops/W) grows exponentially as the feature dimension decreases (the most important driving force to reduce overall power for a given IT workload). Nevertheless, as the feature dimension is approaching 1 nm, the classical scaling suggested by Moore's law, must be coming to an end as quantum effects become pronounced at the atomic scale. In particular Dennard's scaling ceased to hold around 2005 because transistor leakage currents started to mar the simple link between feature dimension and chip performance (not a quantum effect), see also Figure 2.3. At the time, it was predicted that a power wall would be hit; nevertheless, the Green500[8] list indicates that best in class compute performance is still increasing exponentially, two decades after Dennard's scaling died. Indeed, the use of hybrid compute architectures, better communication technology, and AI coordination ensured that compute performance kept on growing.

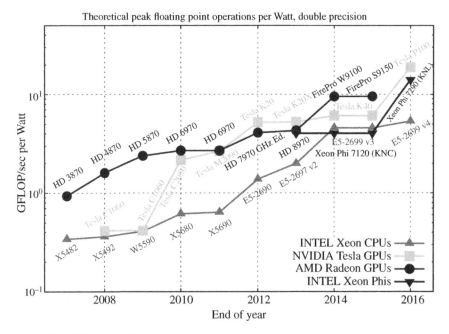

Figure 2.3 GFlops per Watt performance is only slowly increasing. Source: https://www.karlrupp.net/2013/06/cpu-gpu-and-mic-hardware-characteristics-over-time.

---

8 https://www.top500.org/lists/green500.

There are even more avenues for power efficiency gains. Indeed on-chip dynamic voltage and frequency control presents itself as an effective way to actively manage, that is control, the on-chip power consumption. Significant progress toward on-chip power control/management, in particular, for the current multi-core architectures is being made [13] and [14]. In this context, AI methods, such as reinforcement learning, akin to direct adaptive control, are being explored to adaptively assign voltage and frequency control policies for the present workload encountered by the chip [15]. Adaptive, or risk-sensitive stochastic optimal control methods are equally well suited to this task.

Thus far, and despite Moore's law coming to an end, the continued need for more computing power has been met by innovation in new computing architectures. Modern compute units use a hybrid architecture with multiple processing units with diverse compute characteristics, such as classical general purpose central processing units (CPUs), graphic processing units (GPU) with specialized instruction sets particularly well suited for training AI models, reduced instruction set computer (RISC) architectures, and special purpose compute units, for encryption services, or particular AI model execution. An AI-trained model with its own dedicated compute unit can then be designed so as to orchestrate the available workload between the diverse compute units to achieve a particular optimal performance (for a suite of workload conditions). Power management in such environments becomes harder, and benefits from a hierarchical control approach to minimize overall energy and power limitations, as is advocated in [16]. This is another area where AI but also control and systems engineering play a significant role in the computing world, as power and energy remain the key constraints for overall system compute performance. The success of such approaches is measured in the extended battery life in mobile phones, and the continued improvement in GFlop/W performance, even at the highest-performing supercomputer levels such as those listed in the Top500. Data centers benefit equally from these technological developments.

As the age of classic binary computing is approaching its physical limits, the advent of quantum computing indicates that the future of computing remains exciting. The above technical approach of using the available diversity of compute technologies, to improve the GFlop/W performance has a future. However modeling how the various aspects of the workload can be broken up, so that the available compute resources can be orchestrated to achieve a minimum time, or minimum energy objective are becoming combinatorially more complex as the compute paradigm of digital compute is completely different from the quantum computer. Therefore the effective splitting of workload between digital and quantum becomes a really difficult problem, heuristic, and machine learning methods will play a significant role in this space.

2.3.2 Neuromorphic Chips

Because GFlop/W drives data center power researchers and developers are seeking alternatives to the present von Neumann compute architectures, the more so as classical scaling laws (so-called) are nearing their use by date. In particular, neuromorphic chips present an interesting trend. These neuromorphic chips are biomimetic in nature, taking inspiration from the human neuronal brain structure. The architecture of such chips is very different from the classic von Neumann architecture [17]. Neuromorphic chips consist of silicon-based neurons and synapses that process data with ultra-low latency and ultra-low power. For instance, Intel's Pohoiki Beach claims to perform *some* data-crunching and processing tasks up to 1000 times faster than conventional processors like CPUs and GPUs while using only a small fraction of the power [18]. With such improvement in the GFlop/W performance, neuromorphic computing makes an attractive alternative to train and, more importantly to execute AI models where energy efficiency is not compromised despite the massive upscale on the model complexity, see also [19]. Grand initiatives such as the Brain Initiative and Human Brain Project are dedicating tremendous resources in making neuromorphic chip viable as a go-to computing device in the near future. Unlike the traditional computers that run on 0s and 1s, neuromorphic chips use spikes, a set of mixed-digital-analog signals. Reproducible reliability of these spikes remains an issue since they are ultra-sensitive to transistor mismatch and prone to local instabilities. As a result, building large, reliable, IT infrastructure with neuromorphic chips is still an unsolved challenge [20]. This problem also limits the integration of neuromorphic hardware alongside traditional CPUs and GPUs, although specific *sensory* applications are already being pursued. Though the present state of development suggests that in smaller-scale systems, at the edge of the IT network, these neuromorphic systems could be very beneficial already to process and compress images, control robots, and introduce smell and tactile sensors [19]. Given the way these chips represent data internally, it is not clear what are the privacy and security-related implications for spike-based data processing in neuromorphic chips. To build larger scale systems and integrate these meaningfully with the diversity of von Neumann compute systems in play, automatic control is going to play a crucial role. In particular feedback, equally essential in transistor design for von Neumann architectures, is also one of the organizing principles of life, and important to the human brain. Control theory explains why feedback is particularly well suited to improve system level reliability. We can say that building reliable systems out of unreliable components is the raison d'etre of robust control. Nevertheless, our theory and design methods have been developed for treating either digital or analog signals. Dealing with hybrid signals, in particular, the mixed-digital-analog spike signals at the heart of the neuromorphic chips, is hard. The development of robust control, design methods for hybrid signals/systems will be essential to making the neuromorphic chip a computing device of importance in the future of computing, so as to harvest the neuromorphic chip's superior GFlops/W performance. In recent years, a few strides are made toward developing a rigorous robust control theory for stabilizing spiking systems (see [21, 22] for details), and opportunities for further research are endless.

2.3.3 Sleep State Management

Moving back to the classical compute paradigm, apart from control at the chip level, or compute unit level, load assignment across servers is critically important. It introduces yet another layer in the hierarchy of using compute resources for a given workload across the data center. Putting a server in a sleep state consumes far less power than the idle state. In the idle state a compute unit has no meaningful IT workload, but consumes nearly 80% of peak power under full load conditions (see, e.g. [23]).

Therefore controlling the number of servers that are active adds another opportunity to minimize the power consumption. Each server has various levels of sleep state, characterized by the amount of power consumed, and the time it takes to wake up from the sleep state into the idle state, when the unit is able to take on meaningful workloads.

The power benefit from cycling a server from sleep state to idle and back comes at a price. Clearly, there is a delay in availability to take on workload due to the time it takes to wake up the required compute resources. The deeper the sleep state, the more power savings, and the more time it takes to transition into the active state.[9] In order to manage the sleep state, some form of workload buffering and/or re-assignment based on workload priorities is required, and/or a certain performance degradation must be tolerated, which can be priced into the service level agreements.

In its simplest form, adaptive server state management may delay the execution of new workload until the compute resources come available (at the price of requiring buffers of sufficient capacity). A more complex state manager will include a re-prioritization of workloads, provision dynamically, appropriate buffering, and may involve cycling through an enumeration of types of workload to balance overall completion. The most complex sleep control, and most satisfying from an overall power savings point of view, relies on a model to predict the dynamic workload, with the associated prioritization and service level requirements (time

---

9 Typically there are six relevant server states: S0 is the active state (available to take on compute workload), S1–S4 represent the sleep states (from shallow to deep), and S5 simply means that the server is off.

to complete) and a model describing how servers wake up, to provide a dynamic, adaptive scheduler that keeps enough active compute resources to meet the service level requirements of all the workloads. The solution to this problem is known to be a hard (non-polynomial [NP] Hard) problem as explained in [24].

Modeling the diverse workloads is not trivial. Moreover, a lot of work goes into tuning how diverse IT workloads perform[10] well on the available server architecture, or virtual machine or container architecture (which complicates modeling even more). To address the provisioning of the compute resources, various receding horizon control strategies can be considered that take into account the economic pricing of the workload, or the prioritization requirements. Although on a fine grain scale, IT Workload may appear erratic, there are clear patterns across a day (think of web queries and search) and alignment with significant events (think of web services that accompany a festival, or an important sports event). Hence a risk-sensitive approach to achieve sleep state management is important.

Unfortunately, sleep state cycling also affects the serviceable life time of the server, which is much harder to characterize and model. Whereas failures are commonly managed through appropriate redundancy measures, there is here another layer of optimization, where codesign of hardware and its operation play a significant role. As far as we know, this level of control design is not being discussed in the open literature (as yet).

2.3.4 Active State Management

Somewhat more general than sleep state management but only concerned with the active state (and only addressing 20% of the total power budget of a server), is the general problem of virtual machine provisioning so as to minimize the overall power consumption by the physical machines for the workload, under constraints of the quality measures. This is a well-studied problem, see, for example, [25]. It is a nontrivial optimal control problem due to the dynamic nature, and diversity of the workloads, and the combinatorially complex ensemble of all the feasible mappings from virtual machines to physical machines. Due to the underlying problem being combinatorially hard, invariably power schedulers involve a level of heuristics, for which machine learning can be used.

A related problem is the scheduling of containerised[11] IT workloads, as discussed in [26]. This paper is concerned with avoiding server-based powercap limits, and does not present a minimum power scheduler. Software assigned

---

10 Performance metrics may include such things as overall throughput, CPU time as a fraction of overall time, time to complete, total data use, number of network file system requests.
11 Containers or virtual machines represent different paradigms to utilize the same physical server for different workloads.

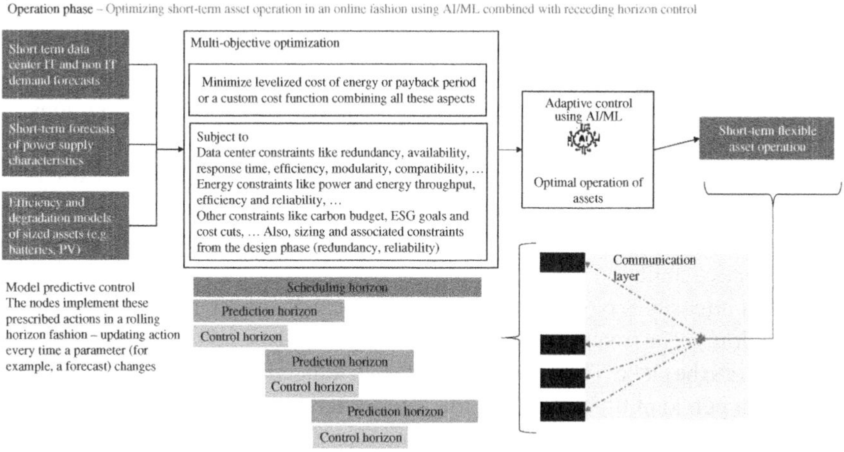

Operation phase – Optimizing short-term asset operation in an online fashion using AI/ML combined with receding horizon control

Figure 2.4 Managing IT workload, sleep state, active state in response to carbon, price or energy signals.

powercaps present a simple and essential power management tool that ensures that the collection of active servers (at any scale of rack, cluster, and center) stay within the available physical power infrastructure at that scale. Analogously, dynamic right-sizing for data centers refers to the practice of adjusting the capacity of a data center to meet changing demands for computing resources. This is typically achieved through the use of power management [27], virtualization technologies [26], such as virtual machines or containers, that allow multiple applications to run on a single physical server [28].

2.3.5 Sleep and Active State Management

Clearly, for a given data center, the combined power management of the sleep states and the active state management of either the virtual machines, or the containers that are running on the physical machines provide the best overall power management approach. Similar approaches can be pursued from a carbon intensity and energy price point of view rather than total power consumption, but the latter is always necessary (Figure 2.4).

2.3.6 Reducing Carbon Footprint

For as long as the world is in transition to zero carbon, there is a carbon benefit to be gained by aligning the IT workload with *greener* energy [29] whenever it is available.

Here we do not consider the option of redesigning or augmenting the data center's supply using renewable energy resources (photovoltaic [PV], wind, battery storage), as that is best approached through codesign (see below). Rather, we consider the question of given periodic measurements of the carbon intensity of electrical supply, how do we align the IT workload to periods of low carbon intensity? This may not be an option for all workloads, certainly not for web queries which are latency-critical, but for many corporate workloads, such as training AI models or periodic reports, for example, delays are acceptable within agreed service limits.

Carbon intensity is typically reported at the local distribution grid level in periodic fashion.[12] The frequency of reporting is determined by the grid operator, and varies across the globe. The carbon intensity will indicate the average carbon intensity over a period of time, based on the actual energy sources participating in the power generation over the reporting window. Over the next few decades, we expect to see the carbon intensity in the electricity grid being reduced to near zero (not addressing the embedded carbon footprint).

Data center operators that have multiple data centers will be able to exploit the different carbon intensities at different data centers and can use the spatial diversity of the compute assets to align workloads so as to reduce the carbon intensity associated with the workload at the cost of some additional communication energy, and within capacity constraints. This control problem is akin to the one discussed above about sleep state management, but simpler, because this works at the scale of an entire data center, and hence the total IT load is less dynamic, as it can be seen as averaged over the diversity of the various IT workloads, and even more importantly, there are fewer choices (fewer data centers than servers) to make. In principle, the control can be implemented by scheduling IT workload preferentially to the data center experiencing lower carbon intensity without violating any center's physical power capacity limits.

A more interesting problem, at a given data center, is to use the information about the carbon intensity to redistribute temporally the execution of the workload, within service level agreement constraints, so as to maximally use the availability of green power. To this end, we need a model for the arrival of new IT workloads, with their expected service priorities as well as a model of the time it takes to complete these workloads with the available compute resources. At the level of a data center, especially when workloads are highly tuned, it is feasible to learn reliable models that predict time to complete for a given resource quantity [30] from measurements. A model that predicts the carbon intensity of the electricity supply over a few hours with some accuracy is also required. Such models are

---

12 See, for example, https://app.electricitymaps.com/zone/SE-SE3 for a graphic representation of carbon intensity in Europe's grid at regular time intervals.

becoming available at the grid operator level, but are not as yet readily available at the distribution level, where in the absence of smart meters at the home that report on PV generation, forecasting solar generation [31] (in particular domestic rooftop PV) remains somewhat elusive. This impacts the veracity of the local carbon intensity signals (which in the absence of domestic PV measurements typically overestimate the carbon intensity in the grid).[13]

Assuming we have a carbon intensity forecast, a scenario-based optimization can then be formulated to minimize the expected carbon intensity of the IT workload by aligning workload with periods of low carbon intensity under the constraint that workloads meet their respective service level requirements [29]. When the diversity of workloads is large, such an optimization can lead to a significant reduction in carbon intensity.

In principle, the same question can be asked for minimizing total energy or indeed cost. Energy price, carbon intensity, and total energy are various aspects of the way IT workloads are executed. These are not automatically aligned. Data center management may be particularly interested in minimizing total energy and total power more particularly to allow more workloads to be executed within the power constraints of the data center's infrastructure. This could be an income maximization strategy, see [32] for example. However, for the clients, it may be particularly important to have a low carbon footprint, a service that may come with a price premium but requires less carbon offsets to be purchased. In general, the most appropriate combination of total energy, carbon intensity, and energy cost will depend on the service level agreements entered into. As the diversity of these service level agreements can be large, the most appropriate criterion for any client may require a special mix of price, energy and carbon intensity signals, and hence different scheduling policies for different clients, which lead to a complex scheduling agent for the data center.

2.4 Codesigning Energy Assets

The cost of building a data center, and all its power components, is very much influenced by the peak power (for IT workload) that may occur under normal operations. Simply assuming that peak power for the data center equals peak power per server times the number of servers will lead to a very wasteful design. The more so as name plate peak power listed on servers severely overestimates the typical peak power under normal workload conditions, especially when workloads are finely tuned to perform well [23]. As all data center components

---

13 See also the start-up www.solstice-ai.com.

come in discrete sizes with discrete power ratings, the way servers form racks, and racks form clusters, with their supporting power, back-up power/energy and cooling units pose a complex and interesting design problem. Moreover, the *best* design will in fact depend on how the data center is going to be operated, and how its workload volume evolves over time. Indeed data centers are constructed in the anticipation of increasing workload. Therefore in the initial stages of deployment there may be a significant overcapacity in power infrastructure, emphasizing the need for more sleep state management to run efficiently overall; also modular power infrastructure will assist greatly. As workload volume increases over time, the need for tight active state load management becomes essential. Modular designs do complicate this picture, as does the improvement in power efficiency with every new generation of server technology.

2.4.1 Formulating a Codesign Problem

A very specific codesign problem is considered here to illustrate the general idea of assets design/sizing taking into account how they will be operated. Because the benefit that may be derived from the assets depends on a non-trivial manner on their size, this design problem is seen as a *codesign* problem. Consider the desire to add a battery to a data center in order to maximize the use of carbon-neutral or low-carbon electricity (for the entire data center or a particular subcollection of servers). The objective is to design a battery that is charged when the carbon intensity of the locally relevant electrical supply is low, and discharged when it is high.[14] The objective is to reduce the carbon intensity of the electricity used in the data center (beyond the background trend in the local grid) at the price of additional embedded carbon footprint and the cost of the battery. The premise is that green(er) energy will create a better market share and/or better profit margins and contribute to the transition to carbon neutrality.

2.4.2 Codesign

Given the battery has an expected asset life of say 10 years, the battery sizing must consider the behavior of the battery over decadal electrical supply scenarios. The scenarios need to reflect the local, carbon intensity, and price at a time scale of wholesale supply pricing (five minutes intervals in Australia) over the next decade. Given the immense optionality at play in the electrical grid market, the locally

---

14 The problem as posed is relevant, as it is a fallacy to think that one can just purchase *green* electricity, and therefore the electrical supply is indeed carbon neutral. Electricity is not provided by a financial instrument!

relevant scenarios[15] will vary widely as well, though given present investments, it is likely that the carbon intensity will decrease over the next few decades, as the world aims to be carbon neutral in some form by 2050.

The economic value, and net present value of the project will depend on the cost to install and maintain the battery, its degradation over time, and in particular, its efficiency in cycling energy through the battery. It will also depend on how the data center can sell its services over that decade, not in the least how valuable the additional reduction in overall carbon intensity for the IT workload is to the clients, who purchase the IT services.

To comprehend how the battery will make a difference to the carbon footprint of the center's operation, we need to consider also an ensemble of scenarios for the expected IT workload, which should realistically reflect the typical dynamic behavior and diversity of the IT workload, with day-to-day, and seasonal variations, but also reflecting the expected growth in demand of the various services over the coming decade, and the pricing structure.

For simplicity sake, when the battery is used to supply the data center, we assume that it supplies the IT workload as well as the corresponding cooling and ancillary services (we are not considering upgrades to this infrastructure over the next decade). The battery must supply the IT workload times the instantaneous PUE (considering the PUE at the same scale as the electrical power carbon intensity and price are being reported). In a first approximation the PUE could be considered to be constant, the averaged PUE over a decade for example. A more realistic approach would use multiple scenarios reflecting that the PUE depends on the environmental temperature and humidity. Given the time span is a decade, climate change as well as the daily variation around climate change (weather) should be taken into account. Earth System Models, augmented with a local weather model to reflect what can be expected at the geolocation of the data center can provide realistic scenarios for weather over a decade, and hence model PUE realistically. Combined earth system and weather models are available commercially, see e.g. IBM's Environmental Intelligence Suite.

Given that battery sizes come in discrete quanta the sizing problem for best return on investment, and/or for a desired carbon intensity reduction is a mixed integer optimization problem. If one also allows for a battery degradation model, it becomes a highly nonlinear mixed integer optimization problem.

In principle, the output of the scenario-based optimization approach is a distribution of optimal battery sizes, and carbon intensity reductions, and derived economic benefits as compared to the no battery, business-as-usual scenario.

---

15 Here scenarios represent samples from a distribution, they are not predictions of the future, rather realistic realizations of a plausible future.

The distribution is over all electricity scenarios, IT workload scenarios and weather/climate scenarios. The ultimate selection has to be made based on an agreed risk profile or return on investment requirement.

The above approach is appropriate only when there is no active control of the workload in response to carbon signals. This serves as the base scenario, without considering the opportunity of flexibility in executing the IT workload.

For any battery size, and a given scenario of carbon intensity in the grid, we should also consider the effect of the control policy for charging/discharging the battery as well as the control of the flexibility that exists in executing the IT workload. Intuitively, the flexibility in the IT workload acts like a virtual battery. Indeed we can delay the consumption of electrical power to benefit from lower carbon intensity. There are constraints in that the electrical energy required for the completion of the workload has to be consumed, and the service level agreements must be met. The battery serves the same purpose but over a different time scale, with different types of constraints. Moreover, the battery comes with an inherent loss in energy, as the available discharge energy is necessarily smaller than the energy required to charge the battery. This is not the case for the active control solution. As commercial batteries achieve almost a 90% cycle efficiency the associated cost is not large. It is the time scale diversity between the battery and the controlled IT workload, as well as the different constraints, that identifies the potential codesign benefit and leads to the true codesign optimization problem. Intuitively, there is an expectation that using codesign, the optimal battery will be (hopefully) significantly smaller than the battery selected, not taking into account the opportunity to re-align workload on the basis of carbon intensity.

The difficulty with taking the workload control into account for the sizing of the batteries is that the complexity of the scenario-based optimization problem grows radically, not in the least because of battery and IT workload management, but also because there are two independent control inputs to work with, each with their own control horizon, and two linked subsystems (battery, and IT workload buffer) with their own sets of constraints.

In Australia, pricing and carbon signals in the electricity market are available at 5 minutes intervals, and the market clears every 15 minutes. The fast market, in which data centers can also play works at even shorter time periods. So, in a 10-year period, there are about 1 M time instances to make control decisions about charging and discharging the battery.

At the operation phase, we consider the short term (order of an hour) forecast of the various IT workloads with their priority levels and time to complete and make a decision, to buffer or wait on the execution, or to purchase electrical supply from the grid (carbon intensity and price), and/or to draw down from the battery

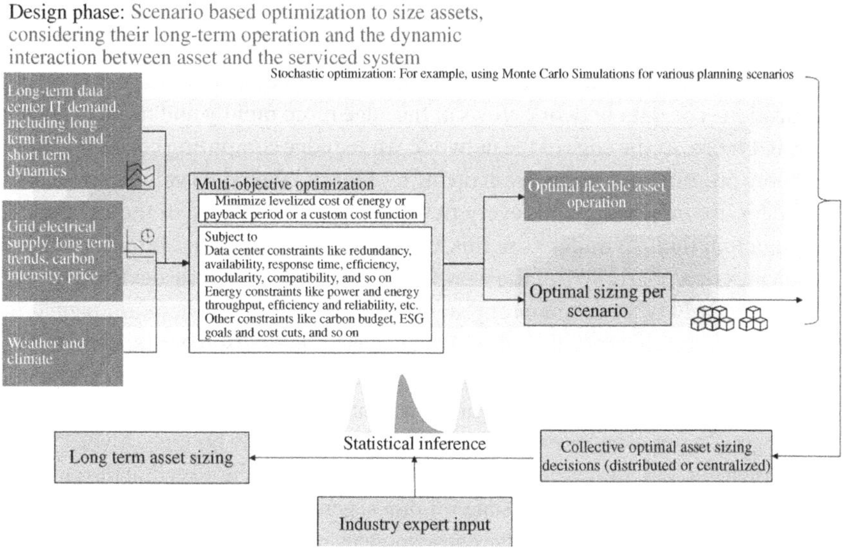

Design phase: Scenario based optimization to size assets, considering their long-term operation and the dynamic interaction between asset and the serviced system

Stochastic optimization: For example, using Monte Carlo Simulations for various planning scenarios

Figure 2.5 Monte Carlo optimization of asset sizing consider the dynamic interactions.

(with its carbon intensity and price, as well as power and energy constraints). The scenarios have to be augmented with a local weather scenario to determine the relevant PUE. At the battery, we need to decide how much to purchase and how much to discharge. Using these scenarios, and for a given scheduling horizon, we solve the receding horizon control problem to minimize the carbon intensity, respecting the constraints that stem from the battery size, battery power limitations, workload service level agreements and workload capacity of the data center. The local optimal controls are implemented, and the receding horizon problem is resolved, until we have completed the 10 year design horizon. The decision process of codesign, with its inputs, and Monte Carlo optimization, is graphically captured in Figure 2.5.

The economic cost of this mode of operation should be compared relative to the alternative like minimizing energy cost, and working without battery.

In [33] a similar problem is considered, not for flexibility in IT workload, but for a different type of electrical demand flexibility, using heat buffers, electric vehicles (EV) charging and the use of PV arrays. It is shown that any flexibility on the demand side is significantly enhanced by adding even a modest battery in the mix, and that the best-sized battery with flexible demand is significantly smaller than the battery that achieves the same benefit without the consideration of flexible demand. We expect this to hold also in the data center situation.

2.5 Beyond the Data Center

The data center is but one (important) aspect of the platform industry supporting digitization. The data network grows at the edge more rapidly and more diversely than at its core. At the edge of the network, we see edge computing becoming more and more prominent, sensors are evolving rapidly and their data volumes, and data flow rates are increasing with every new generation. The edge of the network is very much in flux. To underscore this, consider CISCO's Annual Internet Report. It indicates that nearly 90% of the world population uses a mobile device, and that there are nearly twice as many registered mobile devices as there are people in the world. Presently there are nearly 4 edge devices per person on the planet!? We also have to learn to live with a painful reality that it is often cheaper to keep data rather than to delete them. This is not in a small part due to control engineering, allowing us to put ever-increasing amounts of data reliably on tape drives.[16] This "not forgetting" is actually costing the world, if for no other reason that the amount of junk drowns the useful data. Data mining is getting harder with the second. To make matters worse, think of the consequences of cyber security breaches, and the ease at which fake data and malicious data can be generated and consumed. In contrast, our brain is not like that, forgetting is an important aspect of learning and re-learning, building, and breaking habits, see also [34]. That forgetting is an important aspect of learning, was topical in adaptive control theory in the 80 when a series of papers about forgetting factors were identified as an essential stabilizing influence [35], something we are rediscovering in AI [36], but has been a part of the neuroscience of the brain since Jerzy Konorski coined the term neuroplasticity in 1948. In the next decade we will need to develop a much greater understanding of the systems engineering of data at the scale of the internet. We need to settle a structure for how, where, and when data are gathered, stored, compressed (as in used for modeling) and deleted so that data mining remains meaningful and does not become swamped by data curation issues, privacy and human rights more generally are protected, and to ensure that the promise of more democratic wealth generation through digitization actually comes true.

2.6 Conclusion

The fourth industrial revolution, the digitization of our world, promises optimistically data-driven decision-making at all levels of complexity, at all times,

---

16 FUJIFILM and IBM set a new world record in tape storage in 2020. It was the sixth change in world leading data density since 2006. The new record stands at 317 gigabits per square inch.

for everyone and importantly at speed. Its ubiquity brings with it the inevitable curse that its power can be used to serve noble, ignoble, and outright undesirable objectives. For the present, nothing indicates that its pursuit will realize a world with greater equity, but rather even starker differences between those with access and those with limited access. We should proceed with caution to harvest the benefits without controversies.

Equally clear is that digital enabling technology will consume a great deal of material resources, and energy, all in service of making the rest of the economy more productive. In fact, the evolution and progress in digitization are inevitably limited by its potential to realize productivity in the economy (otherwise the investments will dry up). If digitization truly succeeds, it will be a cornerstone technology to realize a way of living (for all) within the limits of our planet's finite physical resources for as long as the sun shines benignly.

To realize its full potential, systems engineering principles, control and automation at all spatial and temporal scales, from the transistor level to an earth systems level, with an emphasis on codesign for sustainability, are going to play a fundamental role. It is our thesis that this interplay between digitization, and design is a fundamental engineering challenge for the next few decades as we transition to a circular economy.

Aptly, the chapter points to the need to re-consider the data center as a service from a circular economy point of view. This is an immense codesign problem, bringing the technology of all the chapters together as the circular economy requires a much broader system perspective including the mining and manufacturing aspects of all components, as well as their operations, maintenance, re-purposing, and eventually recycling of the entire infrastructure. Potentially this may lead to an entirely re-imagined data, compute and communicate as a service infrastructure on which our society depends. There remains much work to be done, but as gathering data becomes trivial, and data storage becomes too cheap to delete data, and modeling easier, cheaper and more democratically provided enabled by AI tools, we have in principle all the technology to address these challenges. Will we have the political will, the right ethical approach, and earn the social license to proceed?

References

1 Jutta Bolt and Jan Luiten van Zanden. Maddison style estimates of the evolution of the world economy. A new 2020 update, 2020.

2 Joseph E. Stiglitz. The revolution of information economics: The past and the future. Working Paper 23780. http://www.nber.org/papers/w23780, 2017.

3 Iven Mareels, Shally Gupta, Huazhen Fang, Ramneek Kalra, Bozenna Pasik-Duncan, Ralamatha Marimuthu, and Tanishi Naik. Universal access to technology. In *2021 IEEE International Symposium on Technology and Society (ISTAS)*, pages 1–2. IEEE, 2021. doi: 10.1109/ISTAS52410.2021.9629194.

4 Prediction Machines. Updated and Expanded: The Simple Economics of Artificial Intelligence. *Harvard Business Review Press*, 2023. ISBN 9781647824679.

5 Tyna Eloundou, Sam Manning, Pamela Mishkin, and Daniel Rock. GPTs are GPTs: An early look at the labor market impact potential of large language models, 2023.

6 Anders C. Hansen, Matthew J. Colbrook, and Vegard Antun. The difficulty of computing stable and accurate neural networks: On the barriers of deep learning and Smale's 18th problem. *Proceedings of the National Academy of Sciences of the United States of America*, 119(12). doi: 10.1073/pnas.2107151119.

7 European Commission, Joint Research Centre, P. Bertoldi, M. Avgerinou, and L. Castellazzi. Trends in data centre energy consumption under the European code of conduct for data centre energy efficiency. Publications Office, 2017. doi: 10.2760/358256.

8 Srini Bangalore, Bhargs Srivathsan, Arjita Bhan, Andrea Del Miglio, Pankaj Sachdeva, Vijay Sarma, and Raman Sharma. *Investing in the rising data center economy*. McKinsey & Company, 2023.

9 Nestor Maslej, Loredana Fattorini, Erik Brynjolfsson, John Etchemendy, Katrina Ligett, Terah Lyons, James Manyika, Helen Ngo, Juan Carlos Niebles, Vanessa Parli, Yoav Shoham, Russell Wald, Jack Clark, and Raymond Perrault. The AI Index 2023 Annual Report. Institute for Human-Centered AI, Stanford University, Stanford, CA.

10 Richard Lee, Dickon Pinner, Ken Somers, and Sai Tunuguntla. *The case for committing to greener telecom networks*. McKinsey & Company, 2020.

11 F. Brocklehurst. International Review of Energy Efficiency in Data Centres for the Australian Department of Industry, Science, Energy and Resources, 2021.

12 Marco Polverini, Antonio Cianfrani, Shaolei Ren, and Athanasios V. Vasilakos. Thermal-aware scheduling of batch jobs in geographically distributed data centers. *IEEE Transactions on Cloud Computing*, 2(1):71–84, 2014. doi: 10.1109/TCC.2013.2295823.

13 Ryan E. Grant, Michael Levenhagen, Stephen L. Olivier, David DeBonis, Kevin T. Pedretti, and James H. Laros III,. Standardizing power monitoring and control at exascale. *Computer*, 49(10):38–46, 2016. doi: 10.1109/MC.2016.308.

14 Canturk Isci, Alper Buyuktosunoglu, Chen-yong Cher, Pradip Bose, and Margaret Martonosi. An analysis of efficient multi-core global power management policies: Maximizing performance for a given power budget. In *2006 39th*

Annual IEEE/ACM International Symposium on Microarchitecture (MICRO'06), pages 347–358, 2006. doi: 10.1109/MICRO.2006.8.

15 Zhongyuan Tian, Lin Chen, Xiao Li, Jun Feng, and Jiang Xu. Multi-core power management through deep reinforcement learning. In *2021 IEEE International Symposium on Circuits and Systems (ISCAS)*, pages 1–5, 2021. doi: 10.1109/ISCAS51556.2021.9401447.

16 Parth Shah, Ranjal Gautham Shenoy, Vaidyanathan Srinivasan, Pradip Bose, and Alper Buyuktosunoglu. TokenSmart: Distributed, scalable power management in the many-core era. *ACM Transactions on Architecture and Code Optimization*, 20(1), Nov 2022. ISSN 1544–3566. doi: 10.1145/3559762.

17 Donhee Ham, Hongkun Park, Sungwoo Hwang, and Kinam Kim. Neuromorphic electronics based on copying and pasting the brain. *Nature Electronics*, 4(9):635–644, Sep 2021. ISSN 2520-1131. doi: 10.1038/s41928-021-00646-1.

18 Intel Inc. *Intel scales neuromorphic research system to 100 million neurons.* Press release. https://www.intel.com/content/www/us/en/newsroom/news/intel-scales-neuromorphic-research-system-100-million-neurons.html, 2017.

19 Mike Davies, Andreas Wild, Garrick Orchard, Yulia Sandamirskaya, Gabriel A. Fonseca Guerra, Prasad Joshi, Philipp Plank, and Sumedh R. Risbud. Advancing neuromorphic computing with Loihi: A survey of results and outlook. *Proceedings of the IEEE*, 109(5):911–934, 2021. doi: 10.1109/JPROC.2021.3067593.

20 Steve Furber. Large-scale neuromorphic computing systems. *Journal of Neural Engineering*, 13(5):051001, Aug 2016. doi: 10.1088/1741-2560/13/5/051001.

21 Amritam Das, Thomas Chaffey, and Rodolphe Sepulchre. Oscillations in mixed-feedback systems. *Systems & Control Letters*, 166:105289, 2022. ISSN 0167-6911. doi: 10.1016/j.sysconle.2022.105289. URL https://www.sciencedirect.com/science/article/pii/S0167691122000998.

22 Rodolphe Sepulchre. Spiking Control Systems. arXiv e-prints, art. arXiv:2112.03565, December 2021. doi: 10.48550/arXiv.2112.03565.

23 Xiaobo Fan, Wolf-Dietrich Weber, and Luiz Andre Barroso. Power provisioning for a warehouse-sized computer. *ACM SIGARCH Computer Architecture News*, 35(2):13–23, Jun 2007. ISSN 0163-5964. doi: 10.1145/1273440.1250665.

24 Ching-Chi Lin, Pangfeng Liu, and Jan-Jan Wu. Energy-aware virtual machine dynamic provision and scheduling for cloud computing. In *2011 IEEE 4th International Conference on Cloud Computing*, pages 736–737, 2011. doi: 10.1109/CLOUD.2011.94.

25 Zhihua Li, Xinrong Yu, Lei Yu, Shujie Guo, and Victor Chang. Energy-efficient and quality-aware VM consolidation method. *Future Generation Computer Systems*, 102:789–809, 2020. ISSN 0167-739X. doi: 10.1016/j.future.2019.08.004. URL https://www.sciencedirect.com/science/article/pii/S0167739X18324713.

26 Hemant Kumar Mehta, Paul Harvey, Omer Rana, Rajkumar Buyya, and Blesson Varghese. WattsApp: Power-aware container scheduling. In *2020 IEEE/ACM 13th International Conference on Utility and Cloud Computing (UCC)*, pages 79–90, 2020. doi: 10.1109/UCC48980.2020.00027.

27 Minghong Lin, Adam Wierman, Lachlan L. H. Andrew, and Eno Thereska. Dynamic right-sizing for power-proportional data centers. In *2011 Proceedings IEEE INFOCOM*, pages 1098–1106, 2011. doi: 10.1109/INFOCOM. 2011.5934885.

28 Fahimeh Farahnakian, Pasi Liljeberg, and Juha Plosila. Energy-efficient virtual machines consolidation in cloud data centers using reinforcement learning. In *2014 22nd Euromicro International Conference on Parallel, Distributed, and Network-Based Processing*, pages 500–507, 2014. doi: 10.1109/PDP. 2014.109.

29 Priyanka Mary Mammen, Noman Bashir, Ramachandra Kolluri, Eun Kung Lee, and Prashant Shenoy. CUFF: A configurable uncertainty-driven forecasting framework for green AI clusters. In *ACM e-Energy 2023*, 2023. doi: to be published.

30 Tahseen Khan, Wenhong Tian, Shashikant Ilager, and Rajkumar Buyya. Workload forecasting and energy state estimation in cloud data centres: ML-centric approach. *Future Generation Computer Systems*, 128:320–332, 2022. ISSN 0167-739X. doi: 10.1016/j.future.2021.10.019. URL https://www.sciencedirect .com/science/article/pii/S0167739X21004155.

31 J. Antonanzas, N. Osorio, R. Escobar, R. Urraca, F. J. Martinez de Pison, and F. Antonanzas-Torres. Review of photovoltaic power forecasting. *Solar Energy*, 136:78–111, 2016. ISSN 0038-092X. doi: 10.1016/j.solener.2016.06.069. URL https://www.sciencedirect.com/science/article/pii/S0038092X1630250X.

32 Agostino Forestiero, Carlo Mastroianni, Michela Meo, Giuseppe Papuzzo, and Mehdi Sheikhalishahi. Hierarchical approach for efficient workload management in geo-distributed data centers. *IEEE Transactions on Green Communications and Networking*, 1(1):97–111, 2017. doi: 10.1109/TGCN. 2016.2603586.

33 Ramachandra Rao Kolluri, Arun Vishwanath, and Iven Mareels. On the benefits of demand flexibility in the last mile of the grid. In *2021 3rd International Conference on Electrical Engineering and Control Technologies (CEECT)*, pages 180–186, 2021. doi: 10.1109/CEECT53198.2021.9672665.

34 Rodrgo Quian Quiroga. *The forgetting machine*. BenBella Books Inc., Texas, USA, 2017. ISBN 978-1944648-54-1.

35 R. L. Lozano. Independent tracking and regulation adaptive control with forgetting factor. *Automatica*, 18(4):455–459, 1982. ISSN 0005-1098. doi: 10.1016/0005-1098(82)90073-5. URL https://www.sciencedirect.com/science/article/pii/0005109882900735.

36 C. Beierle and I. J. Timm. Intentional forgetting: An emerging field in AI and beyond. *KI-Künstliche Intelligenz*, 33:5–8, 2019. doi: 10.1007/s13218-018-00574-x.

3

Building Automation

Scott A. Bortoff[1], Bryan Eisenhower[2], Veronica Adetola[3], and Zheng O'Neill[4]

[1]*Mitsubishi Electric Research Laboratories, Cambridge, MA, USA*
[2]*Carrier Corporation, Palm Beach Gardens, FL, USA*
[3]*Pacific Northwest National Laboratory, Richland, WA, USA*
[4]*Department of Mechanical Engineering, Texas A&M University, College Station, TX, USA*

3.1 Introduction

Building automation is commonly defined as the automatic centralized control of a building's heating, ventilation and air conditioning (HVAC), electrical, lighting, shading, access control, elevators and escalators, security systems, and other interrelated systems through a Building Management System (BMS).[1] The primary purpose of a BMS is to automate certain aspects of these systems' operations, reducing or eliminating the need for manual labor. The intentions are to reduce energy consumption, to increase security, to improve occupant comfort, to record information for maintenance or accountability reasons, and to provide remote access to this information.

In this chapter, we pose the question: What are the control research challenges that, if addressed, will enable and drive meaningful innovation in building automation in the twenty-first century? We focus on innovations that might mitigate the impact that buildings have on the climate, especially greenhouse gas production. Buildings, as end users of various forms of energy (mainly electricity, natural gas, and oil), account for approximately one-third of greenhouse gas emissions worldwide. Building automation systems (BASs) therefore have a large potential to directly reduce emissions, through improved conservation and efficiency measures, and also to indirectly reduce or eliminate emissions, by enabling behaviors such as demand response (DR), for example, that in turn enable innovations in the supply of renewable electricity.

---

1 https://en.wikipedia.org/wiki/Building_automation.

The Impact of Automatic Control Research on Industrial Innovation: Enabling a Sustainable Future,
First Edition. Edited by Silvia Mastellone and Alex van Delft.
© 2024 The Institute of Electrical and Electronics Engineers, Inc. Published 2024 by John Wiley & Sons, Inc.

In this chapter, we limit our scope to building automation as it pertains to HVAC systems and associated building systems such as active shading, photovoltaics (PVs) and electric grid-interaction, for two reasons. First, HVAC systems are the largest consumers of energy in commercial and residential buildings [1, 2]. Second, performance and behavior of HVAC systems are strongly affected by control, whereas other systems associated with a BMS, such as security or access control, are information systems that are generally outside the influence of control, and/or have relatively little impact on energy consumption. We also limit discussion to common types of residential and commercial buildings, including single or multi-family residential buildings, apartments, office buildings, schools, hospitals, hotels, restaurants, retail buildings and the like. Some types of buildings have specialized energy or HVAC requirements, such as those used for industrial manufacturing, and may fall outside our scope. Other types of buildings such as supermarkets have significant amounts of refrigeration equipment, which makes use of the same vapor compression principles as HVAC. However, for simplicity of exposition, we ignore the specialized needs and research challenges associated with these buildings, although some of the issues we raise, and some of the needs for innovation that we describe, may be applicable to these types of buildings as well.

On the other hand, we adopt a broader definition of building automation than the common one provided above, to include all aspects of the operation, automation, and control of HVAC systems and equipment (and related systems) within buildings, not only a centralized BMS, which typically provides only a supervisory level of control. This is also for two reasons. First, the overall system performance is affected by all levels of control, and nontrivial challenges and opportunities exist even at the lowest level. Second, HVAC systems manifest a very broad variety of physical manifestations, depending on building type and location. Some are so-called "built up" systems, which are designed and constructed as a custom system for a specific, individual building. Other buildings employ factory-built, packaged unitary, or split equipment, where the control function is inserted into the product at the factory, designed to be commissioned and integrated into a broad variety of building types. The control of this type of equipment increasingly includes functionality that is conventionally associated with a BMS, blurring the hierarchical organization. So we consider all levels of building HVAC control in buildings.

This chapter is organized as follows. In Section 3.2, we provide a brief description of a typical vapor compression cycle, which is by far the most common and efficient means to move heat in buildings. We then describe several common types of HVAC systems used in commercial and residential buildings, which are typically organized into a hierarchical structure ranging from equipment at the lowest level to the building or a collection of buildings at the highest. We then describe dynamic models that are used for model-based control development. In

Section 3.3, we discuss industry trends and drivers of innovation, followed by Section 3.4, which articulates their implications on control. In Section 3.5, we describe opportunities for innovation and control research from the need-driven point of view of the industry. In Section 3.6, we imagine a set of vision-driven opportunities for innovation and control research. We conclude the chapter in Section 3.7, returning to the question: How can the control research community impact building automation?

3.2 HVAC Background

Building HVAC systems provide two functions to buildings: (i) Regulation of thermal comfort, typically by regulating indoor air temperature and possibly humidity, and/or regulating certain building construction temperatures, and (ii) ventilation, by actively exchanging some of the inside air with outside air. (Except for extreme climates, residential HVAC systems typically provides only (i), with (ii) being passively achieved by air infiltration through the building envelope.) In order to meet (i), heat needs to be moved, usually from a colder fluid to warmer fluid, requiring energy to be consumed. In order to meet (ii), air must be moved from one location to another. The basic technologies for both has remained the same for more than a century: the vapor compression cycle and fans.

3.2.1 Vapor Compression Cycle

At the heart of nearly every building HVAC system is a closed vapor compression cycle, diagrammed in Figure 3.1, whose purpose is to move heat from a lower temperature fluid to a higher temperature fluid using mechanical (or in some cases

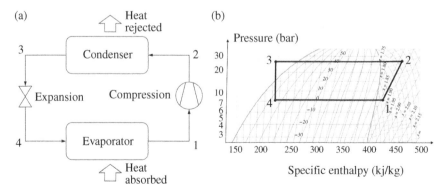

Figure 3.1 Basic vapor compression cycle (a) with corresponding pressure-enthalpy chart (b).

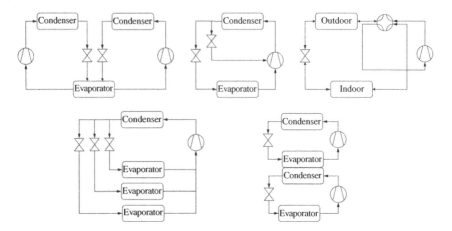

Figure 3.2 Nonexhaustive examples of different vapor compression system architectures (clockwise); dual circuit, economized, reversible (heat pump), multi-evaporator, and two-stage.

chemical) input power. Other physical processes exist for this purpose, such as the thermoelectric effect or magnetic refrigeration, but neither comes close to the energy efficiency and scalability of vapor compression. The basic vapor compression cycle operates as follows. A compressor compresses the refrigerant to hot, superheated gas, which flows into the condenser heat exchanger (HEX), where it condenses, releasing heat to the first fluid stream. Upon leaving the condenser HEX, the refrigerant is expanded through an expansion device, where its pressure (and temperature) drops, but, to first approximation, no heat transfer occurs. At the exit of the expansion device, the refrigerant is typically two-phase (part gas and part liquid), and it enters the evaporator HEX, where it evaporates, absorbing heat from the second fluid stream. Usually, it evaporates entirely to a slightly superheated gas at the exit of the evaporator HEX, where it is sucked into the compressor, completing the cycle. Thermodynamically the cycle is diagrammed using the pressure-enthalpy chart on Figure 3.1b, with numbers corresponding to points between the major components.

A practical vapor compression system usually includes additional components, such as accumulators or receivers, which store liquid refrigerant at various locations in the cycle, valves to change the direction of the refrigerant flow, so that the system can move heat in either direction, and additional HEXs to improve thermodynamic efficiency, and may also incorporate multiple compressors, or HEXs on either side of the cycle, as shown in Figure 3.2.

3.2.2 Fans

Fans remain the basic means of air transport in buildings, and come in a variety of types, including centrifugal, axial, and cross-flow (roll). Although little has changed in the basic principles of operation during the last century, modern fan design and control are increasingly complex, with energy efficiency, operating envelope (due to variable-speed actuation) and acoustic noise all being important design factors, and coupling and dynamic interaction among components of an overall system affecting its controlled operation.

3.2.3 HVAC System Architectures

Building HVAC systems may be classified depending on the building type, size, and location. At the largest end of the spectrum, a large commercial building (or campus) typically will have a so-called "built-up" Variable Air Volume (VAV) HVAC system, consisting of a chilled water plant (chillers, water distribution, cooling towers), air handler units (supply and return air fans, heating and cooling HEXs, filters), air distribution ducts, and terminal units with controlled dampers for each building zone, as shown in Figure 3.3. The chiller uses a vapor compression cycle to move heat from the chilled water loop to the higher temperature condenser water loop. What characterizes this type of system is its use of water as a means to transport heat over large distances, and the ducted distribution of

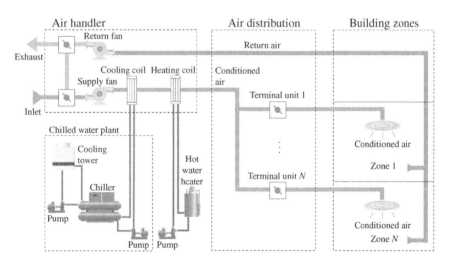

Figure 3.3 Built-up system.

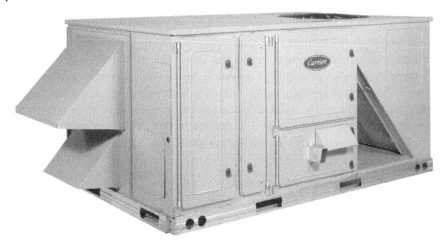

Figure 3.4 A packaged unitary HVAC product. Source: Carrier.

conditioned air throughout the building. This type of system is custom designed for the building and assembled on-site from factory-made components.

Smaller commercial buildings are typically served by so-called packaged unitary HVAC equipment, in which the function of conditioning the air is packaged into a factory-built unit, often placed on the roof of the building. An example is shown in Figure 3.4. Packaged unitary equipment will typically include supply and return fans for ventilation, heat exchangers, a vapor compression system, and possibly gas or electric heat, depending on the application. Usually the vapor compression system is air-source, meaning heat is exchanged directly with the outside air on one side of the vapor compression cycle, while the other conditions the indoor air stream, so refrigerant is used to move heat directly from the indoor air to the outdoor air. (These systems are sometimes called "direct expansion," or DX units.) As in Figure 3.3, the conditioned air is distributed throughout the building, and local zone conditions are regulated by feedback by actuating a variable-position damper in a terminal unit.

A third architecture, commonly used in Europe and Asia, is based on so-called variable flow refrigerant (VRF) equipment. As shown in Figure 3.5, the vapor compression cycle is physically split into two or more separate units (so these are sometimes called "split systems"). The outdoor unit contains the compressor, a heat exchanger, variable speed fans and valves, and possibly refrigerant expansion devices. Indoor units are located throughout the building and contain a heat exchanger, a fan and possibly refrigerant expansion devices. Pipes connect the indoor and outdoor units, and in some systems, additional equipment such as branch controllers are used between the two. These systems usually have variable

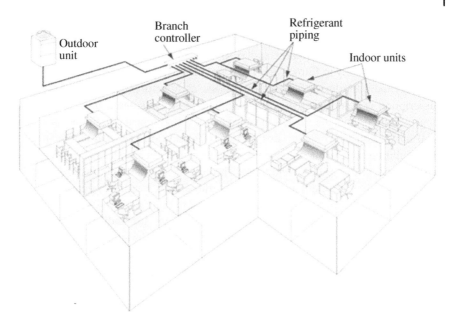

Figure 3.5 A VRF system.

speed compressors, so that the cooling / heating capacity (and the refrigerant flow rate) can be varied. Ventilation is supplied via an Outside Air Processor (OAP) or Dedicated Outdoor Air System (DOAS), which is a separate unit that is similar to a packaged unitary HVAC system, but smaller because it conditions only the outside air that is introduced into the building. These may incorporate energy exchange between the supply and exhaust air, such as desiccant wheels, for energy saving. What characterizes these systems is that they transport heat through the building using refrigerant (instead of water), and they regulate zone temperatures by varying heat flux from each indoor unit. Therefore VRF systems do not require most of the air ducting used in VAV systems and are sometimes called "duct-free." VRF systems have an advantage that in some configurations, they can move heat from one building zone to another (e.g., moving heat from one room in cooling mode to another in heating mode), which is thermodynamically more efficient than exchanging both with the outside air.

For single-family residential building application, many systems in North America are so-called forced-air architectures with a split vapor compression system. The outdoor unit includes the compressor while the indoor HEX is mounted in a supply air duct. Depending on climate, this may be integrated into a gas-burning furnace, which provides heating in winter. In some systems, the vapor compression system can be driven as a heat pump, reducing the need for a furnace. Hydronic systems, which heat water using a vapor compression cycle and distribute it among indoor HEXs, are used for heating throughout much of Europe and Asia. Similarly, split VRF systems are also used for residential application, mainly in Europe and Asia, and increasingly in North America.

A myriad variations of these three basic architectures exist and this short description is nonexhaustive and incomplete [3, 4]. In some commercial buildings, terminal units incorporate heat exchangers that reheat conditioned air. Fan coil units, which include an air-to-water HEX and fan, and modulate heat flux to a building zone using either a water valve or the fan speed. In these systems, chilled and heated water is circulated throughout the building, instead of only the air handlers. In many parts of the world, district heating systems circulate warm water or steam through a piping network to buildings, which then contain their own hydronic heating system. Some buildings use radiant heating and cooling, in which heated or chilled water is circulated through parts of building constructions (usually the floor or ceiling), and heat is transferred convectively and radiatively from these surfaces to building occupants. Hybrid VRF systems are a mix of split VRF systems and hybrid systems; the vapor compression cycle is split, and indoor HEXs exchange heat with a water loop, which distributes heat among a set of terminal units, usually fan coils. Many buildings include a mix of these different architectures, because over the building lifecycle, heating and cooling requirements evolve. For example, computer servers rooms require cooling year round, with high heat flux requirements. Often this is met using a dedicated VRF system even though the building may be an entirely different architecture.

3.2.4 Control

Regardless of the architecture, control of these system is hierarchical, resembling a process control application, although the physical manifestations of the control functions vary by HVAC architecture and equipment type. It can be described as having three layers, although this is a somewhat arbitrary partition. At the lowest levels, feedback control loops regulate process variables to set-points and enforce process variable constraints. Many control loops are SISO, with PID compensation if the actuator is continuously variable, or with hysteresis if the actuator is on-off. At the system level, the control issues concern coordination and functional integration, so that components function together as an automated system or subsystem. At this level, start-up and shut-down sequences are realized, some set-points are scheduled, and some protection logic, error handling, and alarm logic is realized. At the highest level, where the conventional BMS sits, building-level issues are automated, and integration and coordination among building-level systems such as fire and security are realized. Building occupancy schedules are defined, resets (set-point changes to minimum-energy settings) are realized, and data is collected, marshaled and communicated for purposes of diagnostics and remote monitoring.

3.2.4.1 Low Level Control

At this level, the primary requirements are regulation, disturbance rejection, and constraint enforcement of process variables. What characterizes this level of the hierarchy is the use of feedback. Although many feedback loops are SISO, the entire HVAC system and building are dynamically coupled and highly interactive, and robust stability and performance over wide operating ranges and in a variety of building applications, as the equipment and building evolve, is the primary challenge. The quintessential example is zone temperature regulation. In one of the simplest and oldest examples of feedback, a wall-mounted thermostat turns on or off the heating or cooling equipment depending on the difference between zone temperature set-point and measured zone temperature. This is still the most common method of comfort control in residential buildings today.

VAV systems have many interacting feedback loops. Each building zone temperature is regulated by actuating a continuously variable damper located inside the zone's terminal unit, often via a PID compensator. Upstream, the duct air pressure is regulated by feedback to the supply air fan. When the temperature increases in a zone, the dampers open to introduce more conditioned air into the zone. This reduces the duct pressure, and the pressure loop responds by increasing the supply air fan speed to maintain the pressure. The ratio of outside air to recycled air may also be regulated by feedback using a CO_2 sensor. If present, the cooling tower return water temperature is regulated by feedback, typically by varying the cooling tower fan speed. These functions are realized in an embedded digital controller, which often controls multiple zones and interfaces to higher levels of control and is configured (tuned) at commissioning time, as the building is being constructed or modified.

Many low-level control functions lie within HVAC equipment, such as chillers or VRF systems, and are designed at the factory. These so-called Product Inserted Controls (PICs) contain proprietary control algorithms that have, over time, become increasingly complex and sophisticated. For example, in some types of chillers, the basic control actuators are the compressor speed, which can be changed continuously, and an electronic expansion valve setting, which is also continuously variable. (Additionally, chillers that use a centrifugal compressor also control aspects of refrigerant flow into the compressor, usually using a continuously variable inlet guide vane.) If the water pumps are integrated, then their speeds may also be under control, either as continuously variable

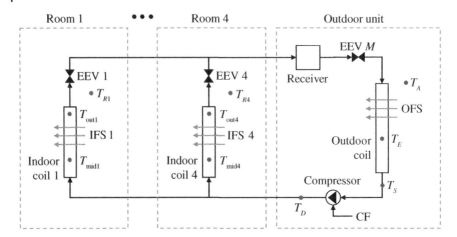

Figure 3.6 Four-zone VRF system showing temperature sensors, indicted with dots, and control actuators, including five electronic expansion valve settings EEV 1, 2, 3, 4, and M, and the compressor frequency (CF). Refrigerant flows clockwise in heating mode and counterclockwise in cooling mode, changed by a valve not shown. Source: [5]/IEEE.

or as on-off. The basic process variables to be regulated in the chiller are the leaving water temperature and the evaporator superheat. Constraints on water temperatures, flow rates, refrigerant pressures and various variables related to the compressor, depending on its physical type, need to be enforced, typically by selector and antiwindup logic. The chiller PIC will integrate start up and shut down sequences, which ramp up down and down actuation and close feedback loops, and also various forms of protection logic that prevent equipment damage. It may also include logic to generate alarms, some diagnostics, and functionality to record data for maintenance uses, e.g. runtime, number of start-stop cycles, etc.

The PIC for a Variable Refrigerant Flow (VRF) system regulates and enforces constraints on internal process variables, and additionally provides for a significant amount of building-level functionality, because as a system, the factory-built VRF provides many of the functions of the building HVAC. For example, consider the four-zone House Air Conditioner (HAC) system diagrammed in Figure 3.6. Even though this is a relatively small VRF system, it has 10 continuously variable control actuators: The compressor frequency, the commanded settings for five electronic expansion valves (EEVs), four indoor fan speeds, and the outdoor unit fan speed. It can run in either heating or cooling mode. It regulates room temperatures at the building level, allowing for different zone temperatures, and also regulates a set of internal process variables while also enforcing several process constraints using a multivariable PID with antiwindup and selector and override logic [6].

3.2.4.2 System-Level Control

At the system level, the primary requirements are coordination, functional integration among components and subsystems, and automation. At this level, the operation of separate components is coordinated in order to achieve a set of system level performance requirements. For example, the chillers, water pumps, valves, and cooling towers all need to work together in order to deliver chilled water to the air handlers. Part of this is to automate start-up and shut-down sequences, which can be realized in a programmable logic controller (PLC) using finite state machines or ladder logic. For example, a large chilled water plant usually includes several different types of chillers, with one being energy efficient across a wide operating range but having relatively small capacity, while the others may be larger and most efficient at their highest capacity. (Typically, the former employ screw-type compressors, while the latter use centrifugal type.) As building heat demand varies, the chillers need to be sequenced in the most energy-efficient manner, so that the larger chillers run at maximum capacity. A supervisory control will encode the logic to sequence the controllers, and will also rotate their individual operation to take into account maintenance issues.

Some set-points for the lower-level feedback loops are determined at the system level, often by a schedule that might depend on measured building conditions or time-of-day and occupancy schedules. For example, the set-point for the leaving chilled water might be scheduled on outside air conditions, with colder set-points corresponding to warmer, humid weather. The colder chilled water will make the HEX colder, which condenses more water, lowering indoor humidity on hot, humid days. So-called resets are included at this level, which modify process variable set-points such as flow rates or supply temperatures to minimum values in order to save energy when the building is in an unoccupied state.

Although feedback loops in the conventional sense are generally considered at the lower level, stability, robustness, and system dynamics are also key issues at this level because HVAC components and subsystems, when interconnected, become dynamically coupled and interactive. Emergent, system level dynamic behaviors can manifest themselves, and, in the limit, do so as undesirable oscillations or outright instability. Furthermore, the building's thermodynamics are tightly coupled to the HVAC equipment's dynamics, and this makes each installation different. Therefore, although most of the design considerations at this level concern automation, significant attention must be paid to overall system dynamics, stability, and robustness.

3.2.4.3 Building Level Control

At the building level of control, control is largely supervisory in nature. The primary issues are set-point scheduling, data collection and remote monitoring, and integration among other information systems such as lighting control, fire and security. Integration of physical systems such as the power supply (electric grid), local electrical generation, energy storage, and the building envelope, are also considered. At this level, system dynamics are less of an issue, and optimization of performance and information integration are the primary considerations.

Zone temperature set-points are determined, usually using a time-of-day schedule, possibly combination with occupancy sensing. Some HVAC set-points such as the supply air or water temperatures may also be determined at this level. A primary objective is performance monitoring, making visible key metrics of building operation, and data analytics are playing an increasingly important role. Some types of diagnostics and fault detection are deployed at this level. Electric power grid integration is also consideration. Building-level energy consumption can be shifted from high-demand periods to lower-demand periods by modifying zone set-points. Demand Response (DR) allows the grid operator to reset zone temperature set-points (e.g. 5 °F for five hours), reducing HVAC power consumption during peak demand periods via communication standards such as Organization for the Advancement of Structured Information Standards (OASIS) Energy Interoperation [7]. Building envelope integration is also considered at this level. Some buildings have active shading or automatically operable windows, which may be actuated by the BMS depending on time of day, weather, etc.

Networking plays a large role at the building-level. Open standards such as BACnet are widely supported and used to collect and disseminate data on standard networking platforms. (However, at the systems level, networking is commonly done using standards that are proprietary to particular manufacturers, with translation provided to open standards at the building level.) BACnet allows for interoperability among a wide range of building automation applications, including HVAC, lighting, fire detection and access control.

3.2.5 Dynamic Models

Dynamic models of HVAC systems are used for several purposes, ranging from equipment-level control design to building-level energy optimization. The physics that govern the behavior of HVAC systems include heat transfer (conductive, advective and radiative) and fluid transport. Very generally speaking, the dynamics can be represented mathematically as a set of hybrid (including both continuous and discrete states) differential and algebraic equations (DAEs),

$$0 = f_x(v(t, t_-), \theta) \tag{3.1a}$$

$$\xi(t) = f_\xi(v(t, t_-), c(t), \theta) \tag{3.1b}$$

$$c(t) = f_c(v(t, t_-)) \tag{3.1c}$$

$$y(t) = h(x(t), z(t)) \tag{3.1d}$$

Here $v(t, t_-) := [\dot{x}(t), x(t), z(t), \xi(t), \xi(t_-)]$ is a vector of continuous and discrete states, $x(t)$ is a vector of differential states (refrigerant pressures and enthalpies, metal temperatures at discrete spatial locations in the vapor compression cycle, water temperatures and flow rates, building construction temperatures, zone air temperatures, pressures and humidities, etc.), $z(t)$ is a vector of algebraic variables, $y(t)$ is a vector of measurements, $d(t)$ represents time-varying boundary conditions or disturbances that are typically not measured, $u(t)$ is a vector of continuous and discrete control actuators, $\xi(t)$ is a vector of discrete states (real, integer and Boolean), which change only at discrete time events t_e and are otherwise constant, $c(t)$ is a vector of Boolean states computed by evaluating relations, e.g. $x_1 < x_2$, which change only at discrete time events t_e and are otherwise constant, and θ is a vector of physical parameters. The subscript t_- denotes time immediately prior to an event at time t, which is needed to define discrete event dynamics. Although the specific structure of (3.1) depends on the HVAC equipment type and architecture, building type and location, use case and modeling assumptions, there are certain general characteristics that are more-or-less universal and are important to consider from a control design and analysis point of view.

3.2.5.1 Large-Scale, Nonlinear, Sparse

Depending on the application, (3.1) tends to be nonlinear and large-scale, with hundreds to perhaps millions of states. This is because heat exchangers are often modeled using finite-volume methods, in which each HEX is divided into a number of segments, with a refrigerant stream, metal walls, and an air (or water) stream associated with each segment. The refrigerant stream can be considered a one-dimensional fluid flow (gas, liquid or two-phase) and energy, mass and momentum balance equations are expressed for each segment. These are coupled to energy balance equations for the air (or water) stream and heat exchange at the metal wall boundaries, and augmented with a set of empirical closure relations describing single and two-phase heat transfer coefficients and frictional pressure drops. Taken together, each HEX can generate several hundred highly coupled, nonlinear differential-algebraic equations, depending on its size, circuiting, and type. Compressors are often modeled with a set of coupled, nonlinear algebraic equations that relate the compressor speed, power consumption and thermofluid boundary states to one another. Expansion valves are also modeled algebraically, as are other components such as water pumps and fans.

These component models are composed into a system model to represent the interconnections among components, and coupled to a set of thermofluid dynamic equations that describe air transport, psychometrics and heat transfer with the building airspace, envelope and outdoor environment. Building envelope constructions are often modeled as layered, with conductive heat transfer between layers, radiative heat transfer among constructions, and convective heat transfer

(a)   3-Dimensional HEX   (b)   R32 density

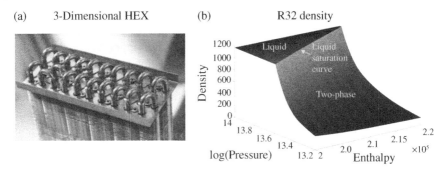

Figure 3.7 Multi-row air-source tube-fin heat exchanger (a), and the density of R32 as a function of enthalpy and pressure (b), showing strong nonlinearity along the liquid saturation curve, which separates liquid (left) from two-phase (right).

to the indoor air. The building model states include the temperature of each solid material layer, along with pressure, temperature and humidity of air in each zone. A fully composed dynamic system model typically has hundreds to tens of thousands of states (the dimension of x), and thousands to perhaps millions of algebraic variables (the dimension of z), depending on the size of the building, HVAC architecture and model resolution assumptions.

Nonlinearity arises in part because of refrigerant and water phase change. The balance equations for the refrigerant stream include partial derivatives of refrigerant density with respect to other fluid variables, which varies by several orders of magnitude depending on the refrigerant state (see Figure 3.7). Heat transfer is a strong nonlinear function of refrigerant state, with significantly higher heat transfer associated with two-phase refrigerant, compared with single-phase. Pressure–flow relationships are also nonlinear. Other nonlinear effects occur because of sensor placement, and will be described in Section 3.5.

Fortunately, composed HVAC system models tend to be sparse: Each equation in (3.1a) depends on a small number of states as a consequence of the finite volume modeling assumptions and the principle of locality. This holds from the small length scales of a HEX up to the large length and volume scales of a building. Figure 3.8 shows the sparsity pattern of the four-zone VRF system diagrammed in Figure 3.6: The Jacobian of (3.1) is less than 1% nonzero. This implies that specialized methods of numerical analysis for causalization, tearing and index reduction [8] are effective to generate control-oriented models and also efficient simulation code.

3.2.5.2 Numerically Stiff

On the other hand, the dynamic system (3.1a) is very numerically stiff. Time constants can range 10 orders of magnitude. The fast dynamics are associated with

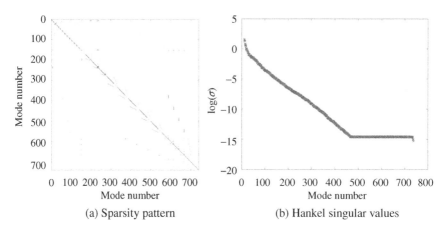

(a) Sparsity pattern (b) Hankel singular values

Figure 3.8 Sparsity pattern of the Jacobian (a) and Hankel singular values (b) for the four-zone VRF system diagrammed in Figure 3.6.

fluid flow in the relatively small heat exchanger volumes, whereas the slow time constants are associated with larger, massive building constructions. Figure 3.9 plots the eigenvalues of the four-zone VRF system diagrammed in Figure 3.6, with the real-axis on a logarithmic scale. The spectrum ranges from about 10 μs to 1 day, or almost 10 orders of magnitude. Numerical stiffness requires the use of implicit numerical solvers such as Differential Algebraic System Solver (DASSL) [9] when computing numerical solutions to (3.1) and can present challenges for model reduction, control, and state estimator design.

It is important to emphasize that the dynamics couple when the HVAC system components are composed into a system model, and this is coupled to a building model. It is incorrect to view the "fluid pressure dynamics" in the vapor compression system as fast. In fact, refrigerant pressure as a state includes both fast and slow modes, and both are excited in normal operation of the system.

3.2.5.3 Hybrid

Finally, (3.1) is a *hybrid* model, containing continuous states x and variables z, and also discrete states ξ. HVAC systems include discrete actuators such as on/off valves which allow for multiple modes of operation. Control systems include discrete event subsystems to realize start-up, shut-down sequences, for example, and other types of logic. Fluid flow models also require discrete states that represent the flow direction; the structure of the continuous part of the equations depends on these states. For some aspects of control design, the operational mode can be considered constant, so that (3.1b) and (3.1c) may be neglected and the plant is treated as purely continuous-time. For the discrete-event aspects of control design, such as

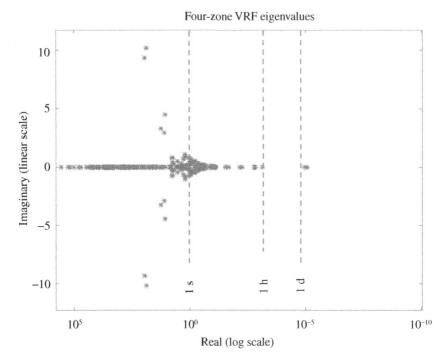

Figure 3.9 Eigenvalues of the Jacobian for the four-zone VRF system diagrammed in Figure 3.6, with real-axis on a logarithmic scale.

the discrete event start-up or shut-down sequences, which are usually represented as finite state machines, (3.1a) might be simplified, and parts of (3.1b) and (3.1c) are synthesized. At some point in the design process, the full hybrid model behavior must be considered, especially when switching among modes. For example, HVAC systems often operate in an on-off cycle when heat loads are low and also switch from heating mode to defrost mode and back. Thus for all aspects and stages of control system design and simulation, multiple models are used to capture the relevant behavior.

3.2.6 CAD Tools

Although (3.1) is a concise mathematical representation of an HVAC dynamic model, it offers little practical usefulness. In practice, Computer-Aided Design (CAD) tools such as EnergyPlus, TRNSYS, Matlab or Modelica are used to construct, manage, analyze (to some limited extent), and simulate HVAC system dynamic models. This is necessary because of their size, complexity and numerical properties.

Particularly noteworthy is open-source *Modelica*[2] [10, 11], which is a computer language for complex, multi-physical, heterogeneous systems such as HVAC in buildings. It is equation-based so that mathematical equations – the language of physics – may be transcribed naturally into the language. It is also object-oriented for organization, so that large models of complex, hierarchical systems, such as building HVAC, may be assembled from libraries of components, such as the open-source Buildings Library [12]. The hierarchical structure of the model mimics the structure of the physical system, and the use of object orientation facilitates reuse, so that component and system libraries accrue through use and grow into valuable business assets. Importantly, libraries of electric power generation and distribution, e.g. PV, are available to meet growing needs to represent these interactions. The language is supported by commercial and open-source tools such as Dymola and OpenModelica [13], which compile a Modelica model into efficient executable simulation code, and can also compute a numerical or symbolic linearization about an equilibrium solution, which is vital for model-based control design.

What separates Modelica from alternatives is its hybrid differential-algebraic model of computation (in contrast, Matlab Simulink has a causal signal-flow model of computation), and the analysis steps that compilers automate to transform an acausal Modelica model into efficient, causalized simulation code. DAEs are more appropriate than ODEs for representing physics, especially compared to causal signal flow models, because they relieve the modeler from the need to make a priori assumptions about causality (which variables affect which other variables), which is error-prone and limits model applicability. Instead, the compiler determines this based on the system structure. This is especially important for thermofluid systems, because the flow direction changes the structure of the equations, and cannot always be determined a priori. The support of hybrid DAEs (meaning a combination of both discrete event logic and continuous variables) enables rigorous implementation of synchronous objects such as flow networks in which fluid flow direction cannot be assumed a priori and also digital control algorithms. The language includes synchronous language features to precisely define and synchronize sampled-data systems, including periodic, nonperiodic, and event-based clocks [14]. A real-time library allows for model interface to real-time input and output [15], allowing for experimental testing of control systems without the need for recoding.

---

2 https://modelica.org.

3.3 Industry Trends and Drivers of Innovation

Several powerful forces are driving the global HVAC industry toward products that are increasingly complex, dynamic, interactive and functionally integrated. Here we summarize those most relevant.

3.3.1 Building-Level Energy Efficiency

Since 1975, ASHRAE Standard 90.1, *Energy Standard for Buildings Except Low-Rise Residential Buildings*, has evolved, defining a minimum acceptable standard for energy-efficient design of buildings (except low-rise) in the United States (Figure 3.10) [16]. The standard covers many building elements including the building envelope, HVAC, hot water, lighting, refrigeration equipment, and on-site electrical generation. This standard bans certain types of energy-inefficient HVAC systems, such as constant air volume systems, while providing guidelines and minimum operating standards for other types of HVAC systems, such as variable air volume (VAV) systems. Most US states apply the standard or equivalent standards for all commercial buildings. Similar building standards exist in Asia and Europe.

3.3.2 Equipment-Level Energy Efficiency and Control

Government agencies rate HVAC systems using data obtained from a small number of fixed thermodynamic conditions and fixed actuator settings, with the

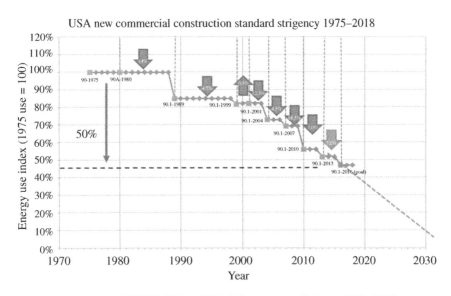

Figure 3.10 Effect of ASHRAE 90.1 on US building energy efficiency, 1970–2018.

JIS B8616 minimum compressor speed test: ramp profile

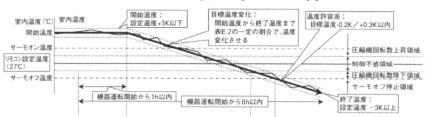

Figure 3.11 JIS B8616 minimum compressor speed test.

control function disabled [17]. This practice dates from the time when HVAC and refrigeration systems used fixed-speed fans and compressors, and fixed-orifice expansion devices such as capillary tubes, so that these measurements accurately represented expected real-world energy performance. Over time, however, manufacturers have introduced variable-speed fans, compressors, and electronically actuated expansion valves, and increasingly complex control algorithms, so that the system can operate efficiently across a broad range of conditions – in principle. Unfortunately, in actual operation with the control function enabled, some variable-capacity systems may provide significantly lower energy efficiency compared to their ratings [18], particularly if little attention is paid to how systems interact within a building. This is sometimes caused by poorly behaved feedback control algorithms.

As a result, there is increasing interest in defining more realistic load-based testing methods to determine the energy ratings for equipment, and the California Energy Commission (CEC), the Canadian Standards Association (CSA) and Japanese rating agencies have all drafted proposed standards that implement such methods [19]. For example, Figure 3.11 illustrates a proposed equipment rating test in which the compressor speed is varied continuously from high to low speed with the control function enabled. New rating tests are no simple matter, because of the wide variety of equipment, and also because a load-based test is considerably more difficult to execute compared to the existing standards, which apply a constant fluid state (temperature, humidity) at the equipment boundaries. Nonetheless, widespread adoption of new efficiency standards and measurement protocols that include the activation of control systems are inevitable.

3.3.3 F-Gas Regulations

Fluorinated refrigerants (F-Gas) are potent greenhouse gasses, accounting for approximately 25% of global warming [20], with recent research showing that they are responsible for half of Arctic climate change in the past 50–60 years [21]. New government regulations are intended to reduce the amount of refrigerant that any one company may sell annually. For example, European F-Gas Regulation 517/2014 is an EU legislative instrument with provisions to reduce F-gas use as

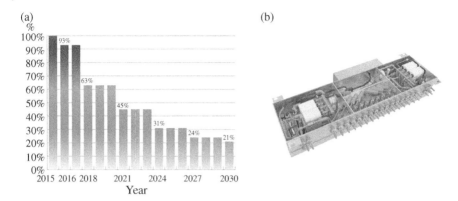

(a)
(b)

Figure 3.12 EU limits on F-Gas use in terms of GWP-equivalent of CO_2 (a), and a Hybrid Variable Refrigerant Flow (HVRF) branch controller, showing heat exchangers, valves, and water pumps (b).

shown in Figure 3.12. Japan and China[3] have implemented, or are implementing, similar laws. This trend, as well as local building codes that incorporate ASHRAE Standard 15 [22], which limits the application of VRF products in some buildings to mitigate health and safety risks of refrigerant leaks, are pressuring all manufacturers to reengineer products to use less refrigerant and/or new refrigerants that have lower global warming potential (GWP).

In response to these trends, many equipment manufacturers have switched the refrigerant used in products from R410A to R32, which has a 3× lower GWP, but is mildly flammable. (Adoption of R32 lags in the United States, largely because of the flammability risks.) In the near future, many of those still using R410A have committed to switching to R-454B, a zeotropic blend of R-32 and R-1234yf that has GWP that is 78% lower than R-410A.[4] (R-1234yf is replacing R-132 in the automotive industry, and has a GWP of less than 1.) This refrigerant presents challenges to engineering efforts because it has temperature glide, meaning the refrigerant temperature changes as a function of pressure in the two-phase region (see Figure 3.1), which makes physics-based models more complex, and may impact control design at the equipment level.

3.3.4 Decarbonization and Electrification

Buildings account for 40% of the primary energy consumption, 74% of the electricity consumption, and 39% of the CO_2 emissions in the United States, with space heating and cooling being large contributors (Figure 3.13). In the United States,

---

3 http://www.env.go.jp/earth/ozone/hiyasu-waza/eng/revised_f-gas_law_in_japan.html.
4 https://en.wikipedia.org/wiki/R-454B.

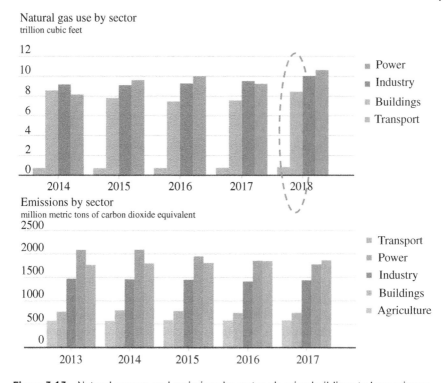

Natural gas use by sector
trillion cubic feet

Emissions by sector
million metric tons of carbon dioxide equivalent

Figure 3.13 Natural gas use and emissions by sector, showing buildings to be a primary target.

over 70 US cities have pledged to become "carbon neutral" by 2050. This is a potent trend, since building codes are governed by state and local municipalities in the United States, rather than the federal government. Several municipalities already have outlawed new natural gas installations and have mandated or incentivized all-electric new construction for some types of buildings [23]. Similar trends are also apparent in Europe and Asia. Electric heat pumps offer the only viable alternative for space heating and cooling. In fact, a Siemens study, commissioned by the city of San Francisco, reports that market adoption of electric heat pumps is the single, most impactful action that the city may take to achieve its objective of an 80% reduction in emissions by 2050 [24].

3.3.5 High-Performance Buildings

High-performance buildings are those which exceed the minimum requirements defined by building codes and regulations, such as ASHRAE 90.1. Zero Energy Buildings (ZEB), in particular, defined as those which consume less energy

Figure 3.14 Mitsubishi Electric's SUSTIE building in Ofuna, Japan.

than they produce on an annual basis, are made possible through advanced building envelope design that maximizes thermal insulation, minimizes undesired air infiltration, minimizes solar fenestration through shading and use of high-performance glass, and may allow for natural ventilation when weather permits. Electric loads must be minimized by using natural daylighting and low energy lighting. Such buildings are often "all electric," using heat pumps for HVAC and hot water, and exploiting waste heat where possible. With sufficiently low power loads, on-site PV generation can meet building needs on average. Significant attention must be paid to functional integration of the building electrical generation, supply, and distribution systems, HVAC, elevators, and lighting. From a control perspective, coordination and functional integration among subsystems, set-point, and schedule optimization, and performance monitoring to make visible key aspects of building energy usage and identify and correct any problems are all critical.

Figure 3.14 shows Mitsubishi Electric's ZEB SUSTIE[5] building, a four-story office building located in Ofuna, Japan, completed in 2020. It was the first building in Japan to be certified as ZEB during construction, and during its first year of operation, achieved its ZEB objective, generating more power than it consumed. The building roof and south-facing window shades are covered with PVs which generate sufficient power to meet demand from HVAC, lighting and elevators. The HVAC system is a highly modular all-electric VRF system, with optimized set-points and sequences. Hot water is generated using a CO_2-based heat pump, and its "waste heat" (actually cold air) is used to help cool the

5 https://www.mitsubishielectric.com/en/about/rd/sustie/index.html.

building in summer. The BMS automatically opens some windows to provide natural ventilation when weather is favorable. All of the component technologies are available commercially.

3.3.6 Renewable Electric Generation

On the energy production side, renewable sources such as wind and PV generate 20% of US electricity, recently overtaking coal [25]. But their temporal variability poses a problem for stable grid operation, and this may limit growth. BASs may offer at least a partial solution. A building has a certain degree of demand flexibility and can reduce or temporally shift some of its thermal energy use without negatively impacting occupant comfort. Demand response allows the grid operator to reduce electric load from buildings by resetting HVAC set-points, providing a means to reduce peak demand, and is useful in extreme weather events. Traditional DR is deployed occasionally in response to price signals or explicit requests from the grid. Today, growing distributed energy resources (such as on-site solar PV and energy storage) and more frequent and extreme weather events are increasing the need and opportunity for integrating buildings into the grid. By taking advantage of building demand flexibility through advanced sensing, communication, and control technologies, it is possible to continuously optimize building operations in response to changing renewable supply levels, dynamic electricity prices, or other grid signals.

In 2019, the United States Department of Energy announced a Grid-interactive Efficient Buildings (GEB) Initiative, which aims to combine energy efficiency and demand flexibility with smart technologies to deliver greater benefits to building owners, occupants, and the electric grid [26]. A follow-on roadmap projected that national adoption of GEBs could yield US$ 100–200 billion in savings to the US power system over the next two decades and 80 million tons annual reduction in CO_2 emission reduction by 2030 [27].

3.4 Consequences and Implications

What are the technical and business consequences of these trends? Firstly, since the 1970s, energy efficiency initiatives have incentivized manufacturers to develop products with continuously variable actuators, replacing on-off components in order to improve efficiency at part load, enabled by the availability of inexpensive inverter technology. Compressors became variable speed.

EEVs replaced fixed-orifice capillary tubes. Variable speed fans and pumps replaced fixed-speed fans and pumps. VAV systems replaced Constant Volume systems.

Of course, all these continuously variable actuators need to be controlled. The problem is, at all levels, ranging from HVAC components up to the building level, these systems are increasingly dynamically coupled, interactive and multivariable. For example, the four-zone VRF system shown in Figure 3.5 has 10 continuously variable actuators and 11 sensed variables. As a dynamical system, it is highly interactive from control inputs to measured outputs. Some VRF systems have hundreds of indoor units and dozens of outdoor units, ganged together.

Secondly, electrification means replacing fossil fuel-burning equipment used for heating (furnaces, boilers, and such) with electric heat pumps (providing a lot of tail wind to the HVAC business). One challenge is to develop heat pumps that operate in more extreme climates. Heat pumps need to operate in colder outdoor air temperatures in winter, and higher ranges of outdoor temperatures in the summer. Moreover, again incentivized by energy efficiency, manufacturers have recognized the need to operate compressors over a wider range of frequencies. While today's heat pump might operate its compressor from about 25 to 125 Hz, tomorrow's will operate from about 5 Hz to perhaps 250 Hz, giving a lower minimum capacity for very low load operation and a higher maximum capacity for high load situations. The lower minimum capacity prevents start-stop operation at low loads, saving energy because the start-up transient draws higher than average amounts of power, while the higher capacity allows for application in increasingly warmer climates.

With the increased range of operation come constraints. The equipment must operate within a set of limits, or it can be damaged. There are upper and lower limits on several process variables, and these must be enforced by the control algorithm. In addition, there are limits to the ranges of actuation. Thus, the challenge here is nonlinearity: The dynamics of a vapor compression system change as a function of operating condition, and constraints must be enforced in operation. This makes the control design problem nonlinear. Gain scheduling and selector logic are often used, but as larger sets of operating conditions are considered, practical design methods may not scale.

Thirdly, HVAC systems are increasingly dynamically coupled with other building systems. Physically, they are coupled to the building's thermofluid dynamics. But HVAC systems are also integrated with other building systems through the system-level and building-level controls. For example, automatically operable windows provide natural ventilation, and this couples the envelope and outside weather to the HVAC system. Demand response and emerging grid-interactive technologies couple the electric supply to the HVAC system. As such, buildings are slowly evolving toward complex systems-of-systems.

3.5 Industry Needs-Driven Innovation

In light of the trends and consequences, what are the industry's needs as it pertains to control? For one thing, precisely what the control community offers: Rigorous, mathematically-oriented model-based design and analysis. Many of today's control challenges can be met by applying mature control theory. However, there do remain some difficulties that, if addressed, will enable more robust products to be developed which, when properly integrated, can move ZEBs from demonstration to mainstream.

3.5.1 Next Generation Modeling and Analysis Tools

The foundations of all popular CAD tools (Matlab, TRNSYS, EnergyPlus, Modelica) were laid in the 1970s–1990s. Although all of these tools proved to be remarkably prescient, and each have evolved in the decades since, certain aspects are getting long-in-the-tooth. For example, Matlab Simulink is very popular and useful for simulating some types of relatively simple HVAC systems, but its signal-flow model of computation requires the user to predetermine causality of variables (the modeler must choose the states, and usually manually express their derivatives in terms of other states), and its numerical solvers simply do not stand up the needs of numerically stiff, nonlinear and large-scale HVAC systems, especially those that include air-source heat exchangers. Modelica overcomes these limits, and is well-suited for the purpose, but the analyses steps such as causalization and tearing are coded into the compilers with the specific intent to generate simulation code, so are largely hidden. It is not possible to access partial results, which might be useful in dynamic analysis. Resulting code is structured for use by state-of-the-art and highly optimized implicit numerical solvers, such as DASSL. However, these are largely single-threaded algorithms that use one processing core (although recent versions allow for some aspects of numerical solution calculation to be parallelized). Finally, the generated code is intended exclusively for simulation use on a desktop computer, requiring a modern operating system.

In terms of next-generation platforms, some of the needs include the following.

1. Symbolic nonlinear model reduction: Because (3.1) can be large-scale and numerically stiff, it would be useful to reduce its dimension and also eliminate fast modes, akin to computing a singular perturbation [28, 29]. How do we compute compute a time-scale separation parameter ϵ as a function of physical parameters, as well as both fast and slow reduced-order models? A designer might want to compute multiple reductions for different values of ϵ, and conduct dynamic analysis on the fast and slow subsystems. This is a mixture of

both symbolic calculus and numerical analysis. Model scale, including a large number of parameters θ makes this a challenge, although sparseness helps.

2. Symbolic Jacobian calculation: The Jacobian of (3.1) is central to application of robust multivariable control theory. Because the systems are large-scale and stiff and the Jacobian is numerically ill-conditioned, it is important to have an exact, symbolic representation available. This is enabled by existing automatic differentiation (AD) technology but is made challenging by the prevalence of highly nonlinear and nonsmooth fluid property functions in (3.1). Some Modelica tools can compute analytic Jacobians, but they require the user to provide some of the component-level derivatives, i.e. the analytic Jacobian calculation is not fully automated.

3. Parallelized implicit solvers: Parallelizable algorithms for numerical calculation of solutions to (3.1), which can exploit multiple cores or GPUs, are needed. Existing algorithms such as DASSL are single-thread. Some aspects of the calculation of numerical solution, specifically the solution of the linear system of equations that lies at the heart of DASSL, can be parallelized, but often this provides little speed up. What else can be parallelized? Building simulation is an especially challenging application because of the stiffness and also long simulation times, e.g. often one year of typical meteorological year (TMY) weather data.

4. Multi-mode systems: Many HVAC systems are multi-mode, meaning subsets of states and equations may become active or inactive as a simulation evolves. For example, if an indoor unit is turned off, it may not be necessary to simulate its dynamics for some use cases. However, existing tools assume the same structure and number of states and equations for an entire simulation. New compiler algorithms are needed to transform acausal descriptions of multi-mode systems into simulation code that allows for a changing dimension of the state and number of equations at run time [30].

5. Embedded code generation: Modelica compilers generate code appropriate for desktop simulation. However, there is a need to take a Modelica representation of a controller and generate code meant for embedded application, which typically lacks operating system support. The developing Embedded Functional Mockup Interface (eFMI)[6] standard aims to close this gap, targeting the automotive industry initially, and there is significant opportunity to mature this technology and target it to the building automation industry.

The developing Modeling-Toolkit.jl [31] written in the Julia language, is a modeling package built on a symbolic computational algebra framework, and may provide a platform for future development. Like Modelica, it allows for users to construct large, acausal models from components, and it is equation-based and

---

6 https://www.efmi-standard.org.

object-oriented for organization. It uses similar algorithms for model analysis and causalization, such as Pantelides algorithm for DAE index reduction. However, Julia and the Modeling-Toolkit allow for a broader use of symbolic analyses, such as automatic differentiation, so that a Jacobian or Hessian of a model may be computed symbolically, ensuring that it is highly accurate when evaluated.

3.5.2 Robust Equipment-Level Control

HVAC equipment is increasingly multivariable, interactive, and nonlinear, needing to operate over increasingly wider operating ranges. Although many feedback control loops are designed as single input single output (SISO), the system as a whole needs to be analyzed as multiple input multiple output (MIMO) using modern multivariable control methods such as disk margins [32, 33]. Often gain scheduling of feedback gains is used to ensure robust stability, but the dimension of scheduling variables can be large. For example, VRF system dynamics change as a function of indoor and outdoor temperature, heat load (or compressor speed), some fan speeds (which are not automatically controlled), and various geometric measurements. Gain scheduling on all of these is cumbersome. In addition, controllers contain selector and override logic to enforce prioritized constraints on process variables, and stability margins need to be guaranteed for all possible combinations of active constraints, in all operational conditions. For these systems, it is common that a constraint is active for long periods of time, and often these situations are considered normal operating modes. In other words, constraints are not to be *avoided*, but are actually designed into normal system operation. It is not uncommon to find an existing constraint-enforcement architecture cannot be extended to handle a new constraint as a design evolves or requirements change. In this case, the entire design might need to be revised, and there is no guarantee that a similar control architecture can be found. The process is time consuming, trial-and-error, and somewhat ad hoc. Finally, designs may be done at a few nominal conditions, but then need to be validated for hundreds to thousands of different conditions to ensure robust performance.

MPC holds some promise. As a design methodology for multivariable systems that can enforce input and output constraints, it seems well-suited to many HVAC equipment-level control problems. In most cases, output constraints can be considered "soft," so that feasibility issues associated with real-time optimization are minimal. But several challenges must be addressed before MPC can succeed conventional selector-logic architectures. First and most importantly, MPC must address robustness with respect to model uncertainty. Robustness margins must be computable using models, so that robust stability can be guaranteed over large variation in the plant and operating conditions [6]. This may require introduction of nonlinear predictive models, so attention to efficient and robust nonlinear model reduction is required. Second, an MPC design needs to consider robust state estimation using available production sensors. This is often overlooked in the literature. Third, hybrid system issues such as start-up and shut-down must be addressed. How is an MPC started up, and modified when parts of an HVAC system are turned on and off? How does the designer deal with different modes of operation? These issues are considered directly by the designer in a selector-type architecture, where a start-up sequence typically closes one loop at a time. Forth, algorithms must be extremely computationally efficient for real-world product use. Significant cost constraints limit the available computing power, requiring memory efficient and computationally efficient solver algorithms. These barriers currently limit MPC to laboratory experiments, but the pressures to develop a robust control methodology that replaces ad hoc selector-logic methods will continue to grow. Each is a challenging research issue in its own right, requiring mathematical rigor to solve, but all require solution for MPC to have a broad application in this industry.

3.6 Vision-Driven Innovation

What opportunities for vision-driven research might lead to meaningful innovation in this industry segment? Here we describe three opportunities: Digital Twins, Model Predictive Control, and Grid-Interactive Buildings.

3.6.1 Digital Twin

A *digital twin* is a set of computer models that serve as a real-time digital counterpart of a physical object or process. The term was coined in the early 2000s in the context of Product Lifecycle Management (PLM) [34] to mean a set of computer representations of a product as it evolves through its lifecycle, from design to manufacture, then to operation, and finally to disposal. The digital twin was envisioned as an electronic repository of all aspects of design, such as 3-D CAD drawings and engineering simulation models, in addition to operational descriptions such as bills of process and operation. It is maintained throughout the product lifecycle via a real-time data stream of measurements obtained from the physical object. It is used to monitor and predict the behavior of the product in operation in its physical environment for diagnostics purposes, or in a variety of interrogative use cases in which future or past scenarios are analyzed to improve the design or operation of a product.

For our purposes, we define an HVAC system digital twin narrowly to be a physics-based simulation model that is combined with measurements and used in real-time operation of the equipment. It may provide a range of benefits such as the following.

- Virtual sensing: Heat flow through a heat exchanger, which is expensive to measure directly, especially for a direct expansion HEX, may be estimated using a model together with a limited set of temperature and actuator measurements. If sufficiently accurate, this may serve as a utility-grade meter for billing purposes.
- Diagnostics: The amount and location of the refrigerant charge inside HVAC equipment, which is difficult to measure directly, may be estimated and used to identify costly refrigerant leaks or conditions that cause refrigerant maldistribution, which can reduce energy efficiency and product reliability.
- Model predictive control: The digital twin model may be integrated into a product-level or building-level MPC, which can command actuator values that optimize a cost function, such as energy use, over a time horizon, and also enforce constraints associated with the equipment or building operation.

For each of these use cases, a dynamic model of the HVAC equipment and possibly the building, is combined with real-time measurements in order to estimate a quantity of interest that is otherwise unavailable or difficult to measure directly. Since dynamic models are used extensively in product development, it seems reasonable to reuse them for this purpose. However, their use in the operational lifecycle phase differs significantly from their use in product development, and the dynamic model must be substantially modified, representing several research challenges.

1. Estimation of states and boundary conditions: From a control theoretic point of view, a digital twin is a state and boundary condition estimation problem: The model initial condition $x(t_0)$ and the disturbance (boundary condition) $d(t)$ must be estimated from feedback of available measurements $y(t)$ and the available model, which can then be used to predict future behavior. But there are challenges to be addressed. First, most state estimation schemes such as the Extended Kalman Filter (EKF) and its variants are formulated in discrete-time, and are structured as a recursive algorithm with a prediction step and a correction step. The correction step modifies the state prediction in order to assimilate measurements. But the model (3.1) includes states and relationships among states that are hard constraints. Most are implicit to the model and are not violated in typical simulations, given a set of physically consistent initial conditions, which is always the case when the model is used in a product development use case. For example, if humidity of the air is considered, there is a maximum limit to the amount of water that air can hold in the gas state at any given pressure and temperature, corresponding to 100% relative humidity.

The model structure and physically consistent initial conditions prevent this constraint from being exceeded in a simulation use case. However, a state estimator might modify one or more states at an update step causing the relative humidity to exceed 100%. This causes the subsequent prediction step to fail numerically, because the model was simply not intended for this nonphysical condition. Indeed, air leaving a cooling coil is usually very near 100% humidity, so this particular example is hardly a corner case. Flow direction is another difficult constraint to handle. The model (3.1a) makes no a priori assumption on flow direction among any of its control volumes (which range from refrigerant tubes to rooms). Flow direction is a computed variable, and the structure of (3.1a) changes as a function of flow direction (downstream fluid states depend on upstream ones). If the update step changes control volume pressures in a way that changes the flow direction, then the model used in the update step is inconsistent with the model used in the prediction step. Again, this is far from a corner case, because pressure drops between control volumes can be very small in normal operation of an HVAC system or in a model of building airflow.

Therefore, the state and boundary condition estimation problem is constrained, with many hard constraints on the state, boundary conditions and parameters that need to be made explicit, although most are not expressed as *explicit* constraints in simulation use cases. Constrained estimation algorithms such as the constrained Ensemble Kalman Filter (EnKF) [35, 36] or optimization-based constrained estimators [37] need to be considered, but even then the designer faces severe challenges. The model (3.1) is large-scale, nonlinear and numerically stiff, so that Jacobians are numerically ill-conditioned. This causes estimation algorithms such as an EKF to have small domains of convergence, making them nonrobust in practice. Data Assimilation methods used in weather forecasting may offer an approach. These problems are also large-scale, nonlinear, and typically include diverse measurements. However, weather models are usually not numerically stiff and do not need to include phase change, compressible fluid flow or widely varying fluid properties to the extent that HVAC models require.

2. Model calibration: Values for elements of the parameter vector θ, including building material parameters and geometry, along with the configuration of the HVAC system, need to be calibrated for every instance. If we consider a digital twin of an entire building and HVAC system, then the number and diversity of parameters can be overwhelming and require a large amount of manual labor, although digital building representations (building information modeling [BIM]) are evolving and in principle can contain much of the needed data. Even if we consider a digital twin of a factory-built HVAC unitary unit, differences in installation details imply that parameters will vary by installation and therefore require calibration.

For digital twins to move beyond one-off demonstrations, calibration will require robust automation, not only to calibrate fixed parameters, but also to configure the model structure. Moreover, some parameters can vary over the lifecycle of a product, and estimating values for these is valuable in order to ascertain the health of the equipment. For example, model (3.1a) is relatively sensitive to the heat transfer coefficients on the air side of a HEX, and these can change over time as the HEX accumulates dirt or corrodes. These may be considered constant for a product development use case, but for a digital twin, an historical evolution is of interest. An important problem is to estimate the amount of refrigerant in a vapor compression cycle. This may be assumed to be constant in a desktop simulation, but slow leaks are common in the field and it is of interest to multiple parties to identify the leak rate and location to facilitate repair. How should an estimator be designed to estimate this slowly varying parameter? Such a problem is deceptively difficult because of the enormous time scales involved, and considering that the refrigerant charge is not a conserved quantity, i.e. neither the numerical algorithm used for integration, nor the model itself, are constructed to conserve charge, so that it might "drift," or change very quickly, during a long-term numerical simulation. The solution to this problem will require reformulation of the model (3.1a), in addition to new types of parameter and state estimation algorithms, and new methods of numerical integration.

3. Tools and platforms: CAD tools used to represent and simulate the model (3.1) are intended for desktop simulation in a product development use case. They do not easily allow for integration of real-time information from measurements, or scaling up of models beyond a single simulation, although most support parametric simulation studies. New tools and platforms need to be developed that will enable development of digital twins beyond desktop simulation, and support eventual deployment. One promising technology for this is the Functional Mockup Interface (FMI),[7] which is a standard for sharing and simulating models created in Modelica. A Modelica model may be compiled into a Functional Mockup Unit (FMU), which is an executable software package that allows the simulation to be executed on a wide variety of platforms, such as in Python, Matlab or Microsoft Excel. Its primary intent is for model sharing, but an FMU allows for two operations that enable realization of constrained estimators. One important feature of the FMI standard is that it depends on only one other standard: ANSI C. This is stable, so that an FMU can be expected to have a long lifetime. This is an important consideration when developing digital twins for HVAC equipment in buildings, which have lifespans of approximately 30 years.

---

7 https://fmi-standard.org.

3.6.2 Building-Level Model Predictive Control

For building-level control, MPC has gained a lot of attention from the research community for more than a decade [38]. At this level of the control hierarchy, economic MPC is technically well-suited to minimization of energy or power consumption or a related cost by recursively solving a real-time optimization problem, which is typically parameterized by HVAC system set-points, subject to a predictive model and constraints associated with human comfort. Many building HVAC MPC strategies have been reported to achieve energy efficiency goals [39, 40] or incorporate demand response objectives [41–44], with benefits of energy savings from 15% to 50%. In contrast to the equipment-level, MPC at this level is less concerned with closed-loop stability and robustness, and more concerned with optimization of a economically meaningful objective function, usually over a longer time horizon of hours to days [6]. (On the other hand, instabilities and oscillations have been observed at this level too [38].) The main benefit of MPC is that it is a systematic methodology for multivariable optimization of energy consumption (or similar economic cost) subject to constraints, which can be nonobvious because of multivariable interactions among systems and subsystems.

Yet, despite its promise and the research efforts to date, transferring MPC into industrial practice remains in its early stages, due to its cumbersome and time-consuming design and implementation procedures (especially building and calibrating predictive models, and tuning of controller parameters) and online computational requirements, etc. [45]. Automating the process of model construction, calibration, and integration with tools that provide for real-time measurements and numerical optimization is needed before MPC at the building level can move beyond demonstration. As it stands, many demonstrations are accomplished with highly skilled but relatively low cost, subsidized student labor. This environment allows for models to be constructed, calibrated, interfaced to data streams and optimizing software for the purpose of conducing a short-term demonstration. But even if this demonstration shows considerable performance improvement relative to some baseline, the process of creating and maintaining the demonstration is not sustainable, nor does it scale beyond the demonstration. If MPC is to be adopted as a building-level methodology, the costs and skill level associated with its deployment must be no greater than what exists today, implying it must be highly automated and result in a robust and transparent (easy to understand, install, debug, etc.) technology. How to do this is a serious intellectual research question.

Many of the barriers to adoption of MPC at the building level are similar to those of a digital twin, which can be considered the predictive model in an MPC algorithm, but the business interests in the digital twin are better aligned. Businesses as end users of HVAC equipment (both building owners and tenants) increasingly need to monitor and account for their energy consumption and CO_2 footprint, for sustainability reasons. Monitoring of energy consumption and continuous improvement to achieve corporate sustainability goals is rapidly becoming a business practice norm, much like financial accounting. This will motivate these parties to invest in development of digital twins. At the same time, equipment manufacturers have strong interest in understanding how their products behave in operation for several reasons: To close the loop in development and improve future products, to deepen relationships with customers, and possibly to develop revenue steam services. As such, multiple parties have aligned incentives to develop digital twin technology, if only for purposes of performance monitoring and offline scenario planning. As it develops, using the digital twin as the core of an automated economic MPC becomes more feasible. Therefore, it is likely that the digital twin will proceed industrial application of MPC at the building level.

One potentially interesting and unexplored research topic is application of MPC over considerably shorter time horizons than are usually considered at the building level. Typically economic MPC time horizons span from many hours to many days, or even longer. However, with solar PV supplying power to a building-scale microgrid, it is possible to consider predictive control over a horizon of minutes, and use video sensors to monitor the sky and predict the PV power supply over a few minute time horizon, which can vary on partly cloudy days. An MPC at this time scale could manipulate the HVAC system (and other building systems such as hot water and lighting) such that its peak (and average) power consumption remains below the available PV supply, which would reduce or eliminate short time-scale demand from the grid. This would require integration of the HVAC equipment control, possibly allowing the building-level control to manipulate constraints in the equipment, such as the maximum compressor speed, which then enforces these constraints locally, instead of manipulating zone set-points. Since the variation in solar load is on a relatively high frequency, the average zone temperature would be minimally impacted, so human comfort would be maintained.

3.6.3 Grid-Interactive Buildings

Beginning in 2019, the US Department of Energy released a series of reports about GEBs that use smart technologies and on-site demand-side resources (DSRs) to provide demand flexibility while co-optimizing for energy cost, grid services, and occupant needs and preferences, in a continuous and integrated way [26, 27]. GEBs can provide demand flexibility via five modes: efficiency, shedding, shifting, modulating, and generation as shown in Figure 3.15.

Energy efficiency and traditional demand response (load shifting) are decades-old strategies intended to reduce the overall energy demand from the grid and curtail energy use during peak demands or emergencies. While

	Load impact	Example measure	Example benefit
Efficiency	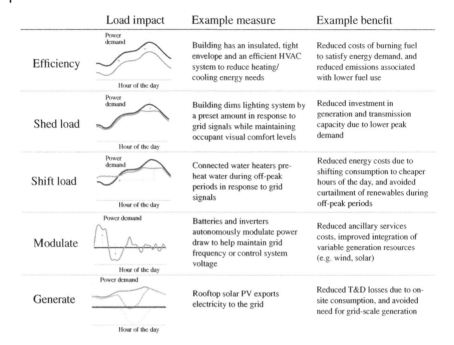	Building has an insulated, tight envelope and an efficient HVAC system to reduce heating/cooling energy needs	Reduced costs of burning fuel to satisfy energy demand, and reduced emissions associated with lower fuel use
Shed load		Building dims lighting system by a preset amount in response to grid signals while maintaining occupant visual comfort levels	Reduced investment in generation and transmission capacity due to lower peak demand
Shift load		Connected water heaters pre-heat water during off-peak periods in response to grid signals	Reduced energy costs due to shifting consumption to cheaper hours of the day, and avoided curtailment of renewables during off-peak periods
Modulate		Batteries and inverters autonomously modulate power draw to help maintain grid frequency or control system voltage	Reduced ancillary services costs, improved integration of variable generation resources (e.g. wind, solar)
Generate		Rooftop solar PV exports electricity to the grid	Reduced T&D losses due to on-site consumption, and avoided need for grid-scale generation

Figure 3.15 Five modes of electric demand flexibility. Source: [27]/U.S. Department of Energy.

these strategies are typically employed separately, to address the rising need for effective demand-side management, buildings must be designed and operated to provide enhanced benefits for buildings' owners, occupants, and the electric grid. Grid-interactive efficient buildings (GEBs) are receiving increased attention in this context for their ability to reduce building energy use and provide grid services through demand flexibility. (A good example of aligned incentives.) The rapid advancement of building technology capabilities and the increasing deployment of behind-the-meter distributed energy resources such as PV provide greater opportunities for buildings to contribute to managing occupant satisfaction and the grid while decreasing the overall cost of building ownership.

Currently, building management operations focus on energy efficiency, cost savings and occupant comfort. Legacy systems do not provide any grid services. Different grid services require varied levels and types of flexibility. For example, balancing generation capacity requires longer load shifts while balancing the grid frequency benefits from fast and short-term changes to usage. The ability to provide more frequent holistic slower services and fast-acting services without compromising building functions differentiate GEBs from what is being done today. Of course, these strategies all require system-level control, which must provide robustly stable performance over wide range of operating conditions. This represents a considerable research opportunity to the controls community, as it has received very little attention as an issue in the development of GEBs.

Buildings can contribute to slower-acting services (e.g. generation capacity relief and contingency reserve) by both lowering the overall need for generation through energy efficiency as well as reducing electricity use for a period (load shed) or changing the timing of electricity use (load shift). The fast-acting services (e.g. frequency regulation) are deployed to correct short-term imbalances in the grid, usually on time scales of seconds to minutes. It requires continuous modulation of building technologies such as battery inverters and variable frequency drives in fans and chillers in response to a signal from the grid operator. Qualified resources must respond to dispatched control signals (within seconds) to increase or decrease their electrical load according to the power grid's needs.

3.6.3.1 Demand Flexibility

Building demand flexibility must be provided within the limits dictated by the building systems and operation. The building demand flexibility is the acceptable modifications to the baseline electricity consumption (via load shed, shift, and modulation) with no undesirable reduction in the building primary functions. Potential methods and metrics for quantifying and predicting building HVAC demand flexibility have been proposed, e.g. [46–48]. Practical methods, in the context GEBs, will need to consider the operation of on-site DERs and account for uncertainty sources affecting building operations. The dynamic aspect of demand flexibility is also vital as building operational efficiency and energy usage vary over time depending on weather conditions, occupancy needs, etc. As systems become interconnected and controlled, closed-loop robustness and stability margins for these strategies must be defined and used for their design and implementation.

3.6.3.2 Control and Coordination Strategies

Broader market adoption of demand flexibility will benefit from scalable control and coordination strategies that can collaboratively engage and aggregate building energy efficiency and demand flexibility technologies, including the enhanced flexibilities offered through on-site thermal or electrical energy storage. While there are several research activities in this direction, a winning solution must be cost-effective, scalable, and easy to deploy with verifiable and sustained benefits. The potential control approaches can be broadly classified into heuristic or rule-based control (RBC) and optimization-based control such as MPC. Although RBC sequences are easier to implement, verify and understand, it is difficult for them to adapt to changing external conditions (e.g. variation in grid peaks) or take anticipatory actions, which are critical to realizing the full benefits of demand flexibility. On the other hand, MPC operates and plans over a time horizon and can optimize multiple objectives. MPC has shown substantial benefits (over RBC) in reducing building operation costs and activating energy flexibility and handling constraints. However, the approach requires system models, forecasting, and nontrivial computational complexities, as previously discussed. Artificial intelligence (AI)-based techniques, such as reinforcement learning, which learn over time by interacting with the environment, have some potential. Still, this technique is relatively immature, doesn't scale well, and hasn't proven effective in real-world building applications. Therefore, there is a clear need for a unified control approach that agglomerates the advantages of MPC and learning-based adaptive techniques with the relative implementation ease and interpretability of RBCs.

3.6.3.3 Cybersecurity in Buildings

Modern BASs require connectivity among systems within the building as well as with outside entities, such as the cloud, to enable low-cost remote management, optimized automation via outsourced cloud analytics, and building-grid integration. As BMSs evolve toward open communication technologies, providing access to BASs through the building's intranet, or even remotely through the Internet, has become a common practice. However, increased connectivity and accessibility come with increased cyber-security threats. BASs were historically developed as closed environments with limited cybersecurity considerations. As a result, BASs in many buildings are vulnerable to cyberattacks that may cause adverse consequences, such as occupant discomfort, excessive energy usage, or unexpected equipment downtime.

The rising demand for enhancing BAS cybersecurity calls for a comprehensive understanding of the BAS cyber landscape. Recent research and studies have been focused on cyber-physical security on BASs, which mainly cover cyberattacks, detection, and defense related topics. But few publications have focused on cyber-secure resilient control strategies specifically for BASs in commercial buildings. Generally speaking, in contrast to other domains that recently received substantial attention such as industrial control and automation systems [49], the security of BASs has been discussed in a less structured manner. An in-depth analysis is still needed to systemically address the cyber-security issues of BASs in the context of the emerging openness and connectivity of intelligent buildings. There are needs for:

- Developing cyber analytics solutions that can minimize the frequency of detection false alarms and accurately diagnose and localize cyberattacks. Preventative strategies are needed as early alarms to catch cyberattacks before they happen on BASs. Solutions that can differentiate cyberattacks from physical faults are also needed to assure targeted response and fast recovery from the effects of adversarial events.

- Developing resilient strategies that can handle multiple simultaneous cyber-attacks and physical faults. Most studies focused on only one type of event at a time. However, multiple cyberattacks and physical faults can occur simultaneously. Therefore, an attractive future direction is developing a flexible detection/defense/control solution to tackle diverse and concurrent cyber threats and faults.

3.7 Conclusions

So, what are the control research challenges that, if addressed, will enable and drive meaningful innovation in building automation in the twenty-first century? Control as a subject (a body of knowledge) is relatively unique in its recognition of robust stability and performance, and of the fundamental and rigorous trade-off between robustness and performance. HVAC systems in particular have slowly evolved from being relatively static systems where closed-loop stability was simply not an issue, into interactive, multivariable, hybrid and nonlinear dynamical systems where robust closed-loop stability is central to correct operation at all levels. This evolution has been driven by a need to improve energy efficiency, and has occurred so gradually over the past 50 years, that the issues of robust stability and performance are still not widely recognized as being critical. (Much like the boiling frog myth, but this is no myth.) From a needs point-of-view, building automation thus offers innumerable opportunities to those knowledgeable in control. These problem-driven issues will only increase in their importance as buildings become more and more electric, dynamically interactive with the electrical grid, and the grid itself becomes more and more decentralized. Robustness needs to be emphasized and researched along side optimization of performance. Directly addressing these problems for real-world products, without changing the assumptions to suit the researcher's constraints or needs, will lead to improved energy and comfort performance of BASs and also a deeper understanding of the gaps and needs in theory. None of these applications involve simply applying existing, well-known methods and tools.

In terms of vision-driven innovation, we have highlighted three areas: Digital Twins, Building-Level MPC, and Grid-Interactive Buildings. These areas represent opportunities for proactive researchers with new ideas and are likely to remain areas of innovation for many decades to come. Combining real-time physics-based models with new algorithms that are needed to estimate unmeasured quantities, some of which are physical and others less so, offer opportunities to those interested in estimation and statistical learning theories. Building level MPC has been researched but has yet to find a strong industrial need, demands a high degree of expertise, and remains largely practiced as case studies. Yet it may offer a means to integrate physical systems (HVAC, grid, for example) as well as models of computation (physics-based models with occupancy and behavior models, e.g. which are learned). There is an opportunity here, but issues related to robustness, often neglected, need to be front and center. Grid-Interaction, and on-site generation, will grow in importance and will increasingly require some attention to dynamic interaction. Existing demand response algorithms, which reduce set-points for a specified period of time, will evolve, and the control field should play a role in shaping this evolution to ensure issues related to dynamic interaction are considered.

Finally, new types of collaborations among partners in academia, industry and government laboratories need to be imagined. Many of these challenges and opportunities are multidisciplinary and nontrivial and are simply not addressable in any meaningful way via the conventional student-constrained, short-term research paradigm common to academia. There is insufficient time for a graduate student to master the required fields and then make a contribution. Longer-term research programs need to be established, with involvement from leaders in industry who are willing to share their knowledge, experience, problems, needs, and challenges, which are often considered proprietary and kept secret, or are poorly understood. Incentives need to be established in academia and at government laboratories for impacting commercial technology and/or industrial practice. (This is especially so for engineering.) On the other hand, industry needs to come together and communicate longer-term needs, opportunities and challenges, and then actively work together with partners in government and academia to gradually solve them. This is not today's funding model. Innovation in this industry evolves slowly, but this does not diminish its importance or compelling nature, and in fact demands new ways of collaborating.

References

1 Azkia Afroz, G. M. Shafiullah, Tania Urmee, and Gary Higgins. Modeling techniques used in building HVAC control systems: A review. *Renewable and Sustainable Energy Reviews*, 83:64–84, 2017.

2 United Nations. Buildings and climate change: Summary for decision - makers. UNEP's Sustainable Buildings and Climate Initiative (SBCI), 2009.

3 ASHRAE. *2020 ASHRAE Handbook: HVAC Applications*. ASHRAE, 2019.

4 ASHRAE. *2020 ASHRAE Handbook: HVAC Systems and Equipment*. ASHRAE, 2020.

5 Scott A. Bortoff, Dan J. Burns, Chris R. Laughman, Hongtao Qiao, Claus Danielson, Abraham Goldsmith, and Stefano Di Cairano. Power optimizing control of multi-zone heat pumps. In *IEEE Conference on Control Technology and Applications*, pages 826–833, Aug 2018.

6 Scott A. Bortoff, Paul Schwerdtner, Claus Danielson, Stefano Di Cairano, and Daniel J. Burns. H-infinity loop-shaped model predictive control with heat pump application. *IEEE Transactions on Control Systems Technology*, 30(5):2188–2203, Sept 2022.

7 David Holmberg. Demand response and standards. *ASHRAE Journal, BACnet® Today & The Smart Grid Supplement*, 53(11):B23–B28, Nov 2011.

8 Francois E. Cellier. *Continuous System Simulation*. Springer, 2006.

9 Kathryn E. Brenan, Stephen L. Cambell, and Linda R. Petzold. *Numerical Solution of Initial-Value Problems in Differential-Algebraic Equations*. SIAM, 1996.

10 Peter Fritzon. *Principles of Object Oriented Modeling and Simulation with Modelica 3.3: A Cyber-Physical Approach*. Wiley, 2015.

11 Modelica Association. Modelica Language Specification Version 3.5. Modelica Association, https://modelica.org/, Feb 2021.

12 Michael Wetter, Wangda Zuo, Thierry S. Nouidui, and Xiufeng Pang. Modelica buildings library. *Journal of Building Performance Simulation*, 7(4):253–270, 2014.

13 Peter Fritzson, Adrian Pop, Karim Abdelhak, Adeel Asghar, Bernhard Bachmann, Willi Braun, Daniel Bouskela, Robert Braun, Lena Buffoni, Francesco Casella, Rodrigo Castro, Rudiger Franke, Dag Fritzson, Mahder Gebremedhin, Andraes Heuermann, Bernt Lie, Alachew Mengist, Lars Mikelsons, Kannan Moudgayla, Lennart Ochel, Arunkumar Palanisamy, Vitalij Ruge, Wladimir Schamai, Martin Sjolund, Berhard Thiele, John Tinnerholm, and Per Ostlund. The OpenModelica integrated environment for modeling, simulation and model-based development. *Modeling, Identification and Control*, 41(4):1890–1328, 2020.

14 Hilding Elmqvist, Martin Otter, and Sven Erik Mattsson. Fundamentals of synchronous control in Modelica. In *Proceedings of the 11th International Modelica Conference*, pages 15–25, 2011.

15 Bernhard Thiele, Thomas Beutlich, Volker Waurich, Martin Sjölund, and Tobias Belmann. Towards a standard-comfort, platform-generic and feature-rich Modelica device drivers library. In *Proceedings of the 12th International Modelica Conference*, pages 713–723, 2017.

16 Bing Liu, Michael Rosenberg, and Rahul Athalye. National Impact of ANSI/ASHRAE/IES Standard 90.1-2016. In *2018 Building Performance Conference and SimBuild*, 2018.

17 Air-Conditioning Heating and Refrigeration Institute (AHRI). *AHRI Standard 210/240. 2017 Standard for Performance Rating of Unitary Air-conditioning and Air-source Heat Pump Equipment*. Air-Conditioning, Heating & Refrigeration Institute (AHRI), 2111 Wilson Blvd., Arlington, VA 22201, USA, 2017.

18 Jeffrey D. Spitler, Laura F. Southard, and Xiaobing Liu. Performance of the HVAC systems at the ASHRAE headquarters building. Technical report, ASHRAE, 2014.

19 Canadian Standards Association. Load-based and climate-specific testing and rating procedures for heat pumps and air conditioners. URL https://publicreview.csa.ca/Home/Details/3313.

20 Donald J. Wuebbles. The role of refrigerants in climate change. *International Journal of Refrigeration*, 17(1):7–17, 1994.

21 L. M. Polvani, M. Previdi, M. R. England, and K. L. Smith. Substantial twentieth-century Arctic warming caused by ozone-depleting substances. *Nature and Climate Change*, 10:130–133, Jan 2020.

22 American Society of Heating and Air-Conditioning Engineers (ASHRAE). *ASHRAE Standard 15-2019 safety standard for refrigeration systems*, 2019.

23 Valerie Volcovici and Nichola Groom. The next target in the climate-change debate: Your gas stove. *Reuters*, 2019. URL https://www.reuters.com/article/us-usa-naturalgas-buildings/the-next-target-in-the-climate-change-debate-your-gas-stove-idUSKCN1VU18Q.

24 Dennis Rodriguez. Reaching 80x50: Technology pathways to a sustainable future. Technical report, Siemens, 2020.

25 U.S. Energy Information Administration, Office of Energy Statistics. U.S. Energy Information Administration monthly energy review, May 2020. URL https://www.eia.gov/totalenergy/data/monthly/pdf/mer.pdf.

26 Monica Neukomm, Valerie Nubbe, and Robert Fares. Grid-interactive efficient buildings. Technical report, U.S. Dept. of Energy; Navigant Consulting, Inc., 2019. doi: 10.2172/1508212

27 Andrew Satchwell, Mary Ann Piette, Aditya Khandekar, Jessica Granderson, Natalie Mims Frick, Ryan Hledik, Ahmad Faruqui, Long Lam, Stephanie Ross, Jesse Cohen, Kitty Wang, Daniela Urigwe, Dan Delurey, Monica Neukomm, and David Nemtzow. A national roadmap for grid-interactive efficient buildings. Technical report, Lawrence Berkeley National Laboratory; The Brattle Group; Energy Solutions; Wedgemere Group; U.S. Dept. of Energy, 2021. URL https://escholarship.org/uc/item/78k303s5.

28 Mathukualli Vidyasagar. *Nonlinear Systems Analysis*: Second Edition. Prentice-Hall, 1993.

29 Hassan K. Khalil. *Nonlinear Systems*: Third Edition. Prentice Hall, 2002.

30 Albert Benveniste, Benoit Caillard, and Mathias Malandain. The mathematical foundations of physical systems modeling languages. Project-Team Hycomes 9334, INRIA, Apr 2020.

31 Yingbo Ma, Ranjan Anantharaman Shashi Gowda, Chris Laughman, Viral Shah, and Chris Rackauckas. ModelingToolkit: A composable graph transformation system for equation-based modeling composable graph transformation system for equation-based modeling. Technical report, Julia Computing, 2021.

32 Kemin Zhou, John Doyle, and Keith Glover. *Robust and Optimal Control*. Prentice Hall, 1996.

33 Peter Seiler, Andrew Packard, and Pascal Gahinet. An introduction to disk margins. *IEEE Control Systems Magazine*, 40(5):78–95, Oct 2020.

34 Michael Grieves. Origins of the digital twin concept. Technical report, ResearchGate, Aug 2016.

35 David J. Albers, Paul-Adrien Blancquart, Matthew E. Levine, Elnaz Esmaeilzadeh Seylabi, and Andrew Stuart. Ensemble Kalman methods with constraints. *Inverse Problems*, 35(9):095007, Aug 2019.

36 Vedang Deshpande, Christopher R. Laughman, Yinbo Ma, and Chris Rackauckas. Constrained smoothers for state estimation of vapor compression cycles. In *Proceedings of the American Control Conference*, 2022.

37 Graham C. Goodwin, María M. Seron, and José A. De Doná. *Constrained Control and Estimation: An Optimization Approach*. Springer, 2005.

38 Ján Drgoňa, Javier Arroyo, Iago Cupeiro Figueroa, David Blum, Krzysztof Arendt, Donghun Kim, Enric Perarnau Ollé, Juraj Oravec, Michael Wetter, Draguna L. Vrabie, et al. All you need to know about model predictive control for buildings. *Annual Reviews in Control*, 50:190–232, 2020.

39 Wei Liang, Rebecca Quinte, Xiaobao Jia, and Jian-Qiao Sun. MPC control for improving energy efficiency of a building air handler for multi-zone vavs. *Building and Environment*, 92:256–268, 2015.

40 Frauke Oldewurtel, Dimitrios Gyalistras, Markus Gwerder, Colin Jones, Alessandra Parisio, Vanessa Stauch, Beat Lehmann, and Manfred Morari. Increasing energy efficiency in building climate control using weather forecasts and model predictive control. In *Clima-RHEVA World Congress*, number CONF, 2010.

41 J. A. Candanedo, V. R. Dehkordi, and M. Stylianou. Model-based predictive control of an ice storage device in a building cooling system. *Applied Energy*, 111:1032–1045, 2013.

42 Sergio Bruno, Giovanni Giannoccaro, and Massimo La Scala. A demand response implementation in tertiary buildings through model predictive control. *IEEE Transactions on Industry Applications*, 55(6):7052–7061, 2019.

43 Luca Fabietti, Tomasz T. Gorecki, Faran A. Qureshi, Altuğ Bitlislioğlu, Ioannis Lymperopoulos, and Colin N. Jones. Experimental implementation of frequency regulation services using commercial buildings. *IEEE Transactions on Smart Grid*, 9(3):1657–1666, 2016.

44 Evangelos Vrettos, Emre C. Kara, Jason MacDonald, Göran Andersson, and Duncan S. Callaway. Experimental demonstration of frequency regulation by commercial buildings— Part I: Modeling and hierarchical control design. *IEEE Transactions on Smart Grid*, 9(4): 3213–3223, 2016.

45 Jiří Cígler, Dimitrios Gyalistras, Jan Široky, V. Tiet, and Lukaš Ferkl. Beyond theory: The challenge of implementing model predictive control in buildings. In *Proceedings of 11th Rehva World Congress*, Clima, volume 250, 2013.

46 Han Li, Zhe Wang, Tianzhen Hong, and Mary Ann Piette. Energy flexibility of residential buildings: A systematic review of characterization and quantification methods and applications. *Advances in Applied Energy*, 3:100054, 2021. ISSN 2666-7924.

47 V. A. Adetola, F. Lin, and H. M. Reeve. Building flexibility estimation and control for grid ancillary services. In *High Performance Buildings Conference*, West Lafayette, IN, 2018. Purdue Herrick Labs.

48 Hong Tang and Shengwei Wang. Energy flexibility quantification of grid-responsive buildings: Energy flexibility index and assessment of their effectiveness for applications. *Energy*, 221:119756, 2021. ISSN 0360-5442.

49 Vitor Graveto, Tiago Cruz, and Paulo Simões. Security of building automation and control systems: Survey and future research directions. *Computers & Security*, 112:102527, 2022.

4

Future Impact and Challenges of Automotive Control

Stefano Di Cairano[1], Carlos Guardiola[2], Andreas A. Malikopoulos[3], and Jason B. Siegel[4]

[1]Mitsubishi Electric Research Laboratories, Cambridge, MA, USA
[2]Escuela Técnica Superior de Ingeniería Industrial, Universitat Politècnica de València, Valencia, Spain
[3]Mechanical Engineering, University of Delaware, Newark, DE, USA
[4]Mechanical Engineering, University of Michigan, Ann Arbor, MI, USA

4.1 Introduction

There are few mechatronic devices as pervasive and as impactful on our life as cars. Every year approximately 100 million new cars are produced, resulting in approximately 1.5 billion cars on the road worldwide. Even though traditionally the main focus on automotive vehicle development has been on the mechanical technology, the engine, the aerodynamics, the chassis, today's vehicles often include more than 100 micro-processors and micro-controllers, the majority of which execute, or a used by, a control system, showing that in fact cars are probably the largest produced type of cyber-physical system.

The increase in on-board computational capabilities and data from sensors, coupled with connectivity to off-board resources, both data and computations, in the infrastructure and in other vehicles, suggest that over the next 10–20 years several significant development in automotive technology will be enabled or even driven by software features that rely extensively on control system technologies. This has been a trend already an ongoing for years, with several of the newly introduced technologies being enabled by control algorithms: from pollutant emission reduction, which requires precise control of gas mixtures and combustion, to vehicle stability features that maintain traction and avoid loss of lateral control, to cruise control features that reduce driver fatigue and speeding risks in long trips, to energy management that minimizes fuel consumption in hybrid electric vehicles (HEV), just to cite a few.

The Impact of Automatic Control Research on Industrial Innovation: Enabling a Sustainable Future,
First Edition. Edited by Silvia Mastellone and Alex van Delft.
© 2024 The Institute of Electrical and Electronics Engineers, Inc. Published 2024 by John Wiley & Sons, Inc.

Several new and improved technologies are expected to appear in vehicles in the next 10–20 years, and in fact the next two decades may offer the most radical changes to automotive vehicles since their introduction. Vehicles are moving at fast pace toward being partially or fully powered by electricity, reducing pollution and fuel consumption, and hence changing radically their architecture in terms of propulsion plant, fueling system, key components and architecture. The next radical change appears to be the role of the driver that, with several assistive and automated driving technologies expected to be introduced, will work in coordination with the control systems in the vehicle, sometimes supported, sometimes delegating, and sometimes supervising. In addition, vehicles may no longer be seen as isolated (closed) systems, as communication between them and with the surrounding infrastructure may enable close coordination, information sharing, and overall optimization of the entire road and transportation network.

While it is impossible to cover all the changes expected in the vehicles in the next two decades, this chapter looks at the role of the control in some of the main anticipated developments in vehicles, specifically in the areas of powertrain, electrification, driver assistance and driving automation, and connectivity and integration with transportation systems. In particular, our objective is to highlight the potentials, the challenges, and the overall role of control system technologies in enabling, and sometimes in providing the foundations to, such developments.

4.2 Internal Combustion Powertrain

4.2.1 A Short Review of the Historical and Future Requirements

For more than a century, automotive applications have been almost exclusively powered by Internal Combustion Engines (ICE). By the end of nineteenth century early automotive industry was eager to adopt and develop the yet-to-be high power density and reliable new technology developed by Nicolaus Otto and Rudolf Diesel, between many other pioneers. Despite the many applications of the ICE, as power generation, marine or industrial, its evolution has been, to a high extent, attached to the demands of the automotive industry. Now, for the first time in history, the automotive ICE is under scrutiny.

A short review of the increasing list of requirements on the ICE, with historical perspective, is provided next.

4.2.1.1 Performance and Fuel Consumption

The first decades following the initial applications were characterized by a continuous improvement of the power density and reliability of the ICE. For that, and between other technologies, the engine was equipped with a set of mechanical, pneumatic, and fluid-dynamic systems able to regulate its different actuators. Notably, supplied fuel was provided by carburetors relying on a set of throttle valves and the Venturi effect, and spark advance adaptation to the engine speed and load was possible by using centrifugal and pneumatic actuated systems. These engine technologies still survive in nonautomotive applications as ICE-powered machine tools.

In addition to performance, fuel consumption concerns sharply emerged following Oil Crisis, as a consequence of the rising prices of oil-derived fuels. Performance and fuel efficiency were the driving forces behind the adoption of technologies such as direct injection (in both gasoline and diesel engines), turbocharging, and variable valve timing. In all these cases, engine control plays a relevant role.

4.2.1.2 Emissions

By the end of 1960 decade, developed societies started to be concerned about air pollution: in the US, Clean Air Act was enacted in 1963, with major modification upon the creation of the Environmental Protection Agency (EPA) in 1970, and first European regulations on motor vehicle emissions date from 1970. The onset of the emissions standards marked a turning point in the ICE evolution, since the ICE was now subject to strict emission levels not to be surpassed.

After-Treatment Systems (ATS), made possible by injection systems and earliest electronic control systems (including feedback control with the lambda sensor), were soon a requisite. From that moment on, engine development has been tightly tied to engine control. Not only because new technologies require a close control for their correct operation but because required On-Board Diagnostics (OBD) functionalities *de facto* rule the use of electronic control systems.

Current emission regulations are founded on three major principles:

1. A continuous decrease of emission limits: Major markets have applied along the years a persistent decrease in the regulated emissions, and included additional species to the emission list. As an example, Figure 4.1 depicts the particulate matter and NOx limits for EU diesel powered passenger cars, where drastic reduction cuts exceeding 90% for NOx and 97% for particulate matter have been applied from Euro 2 to Euro 7 regulation.
2. Emission limits are to be satisfied in realistic use scenarios : While past certification procedures were based on under-demanding driving cycles, latest advances in regulation use more demanding cycles for the type certification, and implement real driving emissions (RDE) testing, where the vehicle is tested in real traffic conditions.
3. Emission limits are to be satisfied along vehicle (and fleet) service life : In addition to the mandatory OBD functions, In-Service Conformity (ISC) procedures

evaluate a sample of the fleet and can trigger an emission-related recall if the level is found to have surpassed a legislated threshold. In the latest proposals as European Union Euro 7, yet to be enacted, On-Board emissions Monitoring (OBM) is introduced as a mean to ensure the vehicle emissions conformity and performance over its lifetime. The main objective will be to detect high emitters, ease their diagnostics and take the appropriate actions by using the information provided by on-board emission sensors.

4.2.1.3 Decarbonization of the ICE

While CO_2 was left out of the earliest emission regulations, this is no longer the case: following decades-long debates, world markets have faced the regulation of the CO_2 emission from the automotive sector. The *decarbonization* of the transport is usually referred to as the *phasing-out* of the ICE, and past progressive limits to CO_2 emission at a fleet level have been replaced by regulations limiting the use of the ICE. Norway, who announced bans for both gasoline and diesel new cars starting from 2025, was soon followed by most major economies, which enacted or announced plans for decarbonizing the passenger car market and fleet in the next few decades.

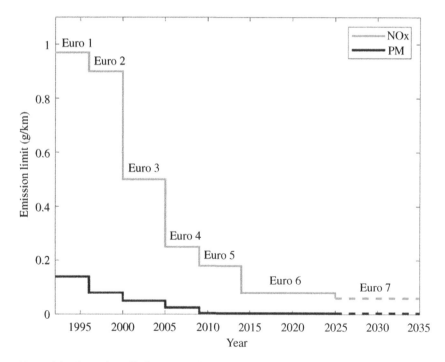

Figure 4.1 Evolution of NOx and particulate limits for diesel passenger cars in the EU.

While the scenario for the ICE in the automotive sector is challenging, some opportunities still exist: in contrast with some future bans of any emitting vehicle, and as a direct consequence of any kind of combustion powertrain, other bans exclusively apply to the fuel (e.g. diesel and gasoline), its carbon content, or origin. In some cases, ICE will still be accepted as a component of *hybrid electric powertrains* (HEV) of various degrees. In many cases, transport fleets are not affected by the ban, or the effect is delayed since no suitable option for long-distance transport exists in the short term.

4.2.2   ICE Technologies and Control Challenges

As it happened in the past with the challenges derived from the emissions regulations, a set of technologies are under development. Arguably, some of these technologies will not reach a sufficient degree of maturity in time, or the manufacturers may opt not to use them because market shrinkage may make them economically unfeasible. In some cases, the final application may be restricted to freight market, where finding a suitable replacement for the ICE is equally challenging.

In Figure 4.2, a keyword co-occurrence map for the papers of the past 10 years is shown for the case of diesel engines. Similarly, interest has been raised toward new fuels (notably biodiesel but also natural gas or dual fuel) and after-treatment (with

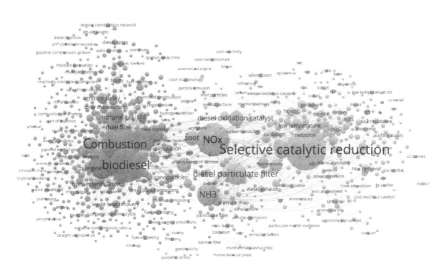

Figure 4.2   Keyword co-occurrence map for papers on diesel engines, by Aronis [1]/ Universitat Politècnica de València after analyzing author keywords of more than 14,000 papers on diesel engines.

a preeminent position of selective catalytic reduction). As it will be highlighted in the next paragraphs, engine control is a transversal key enabler for most of these technologies, continuing the trend of the past decades of a growing importance of engine control.

4.2.2.1 Hybridized ICEs and Electrification

The first mass-produced HEV was marketed in 1997, and now HEV represents a significant share of the total amount of vehicles [2]. In HEVs, the ICE is combined with an energy storage system (battery) and at least one electrical motor. HEVs vary in their degree of electrification ranging from mild HEVs, where the electrical motor is used for assistance (in some cases assisting vehicle acceleration, but in other cases restricted to start-and-stop or for a set of electrically assisted systems), to full hybrids. In the case of plug-in hybrids (PHEV) the ICE is used as a range extender once the batteries are depleted, and the operation of the powertrain is expected to be mostly electrical. Several HEVs architectures exist, such as series, parallel, and series-parallel hybrids [3].

The integration of the ICE into the hybrid architecture opens a series of control problems. On one hand, the ICE operation is no longer defined by the vehicle speed and torque demands: it is now possible to decide the operation as the excess power may be used for charging the battery, and the power defect may be covered by the electrical motor if the HEV layout allows it. In some cases, the controller can also decide the engine speed in a flexible way, if there is no mechanical link between the ICE and the wheels, or if the gearbox allows it. As a consequence, an energy management system (EMS) must be used. The EMS is in charge of deciding the operation of the ICE, considering the torque requirements, the estimated state of charge (SOC) of the battery, and the use conditions.

EMS has been a quite active field of research in the last 20 years. The problem may be easily written as an optimal control problem, which soon replaced initial expert-rule approaches. The problem is usually stated as minimizing the fuel consumption over a trip, subject to a set of constraints on the system operation (fulfilling torque request, not exceeding power rating of the different elements, etc.). Several approaches have been used, ranging from dynamic programming [4], Pontryagin Minimum Principle and its simplification as the Equivalent Consumption Minimization Strategy (ECMS) [5, 6], to model predictive control [7, 8]. Main difficulties are attached to the estimation of the future driving profile [9] and of the battery SOC [10].

On the other hand, ICE operation in an HEV implies a discontinuous operation of the combustion engine. Since the emissions are significantly affected by the engine and after-treatment temperature, it is necessary to take into account the impact of the frequent start-and-stop of the engine. For example, in [11], emissions are added to the EMS control problem.

4.2.2.2 New Fuels and Combustion Concepts

The use of new fuels specifically targets the decarbonization of transport. Attention is moving from biofuels as biodiesel (for diesel engines) or ethanol (for spark-ignited engines), which are currently used in blends with conventional petrol-derived fuels, to low or non-carbon fuels. Fuel variability has a direct effect on engine operation and the adaption of the control settings may be required [12]. Compressed and liquified gas have also gained attention for replacing diesel in heavy-duty applications, or in dual fuel implementations already in the market.

In recent years the use of carbon-free fuels has been under research, with hydrogen and ammonia standing out as the two more feasible alternatives. If implemented, those fuels will allow the ICE to operate without CO_2 emissions, but the "green" production of the fuels is still to be scaled, and both fuels present significant distribution and storage problems. Hydrogen can be easily used in spark-ignited engines, but needs to be compressed and presents a low energy to volume ratio; hydrogen-fueled ICE combustion is not free of NOx production, so specific aftertreatment is to be developed and operated. Ammonia has a better energy-to-volume ratio, but its autoignition temperature is approximately 200°C higher than current fuels [13], creating problems to the ignition process for either spark ignited and compression ignition (with diesel pilot injection in dual-fuel operation), and significant unburnt fuel at the engine exhaust.

Combustion control is an active field of research, especially when the limitations of classic spark-ignited and compression-ignited engines are to be overcome. Following the development of the common rail direct injection diesel engines and the gasoline direct injections, the focus has moved to the use of low temperature combustion harnessing the fuel-controlled autoignition, and the optimization of dual-fuel engines. A number of concepts have been presented as, HCCI as PPCI, and control solutions have been researched usually requiring in-cylinder pressure sensing and making use of additional control inputs for improved control authority, as a secondary fuel injection [14], control on fuel blend for controlling fuel reactivity as in RCCI [15], or controlling the residual fraction using a combination of exhaust gas recirculation (EGR) and negative valve overlap (NVO) [16, 17]. Cycle pressure signal feedback has been used mostly for next-cycle control but also for same-cycle control requiring fast signal processing with a FPGA [18, 19].

4.2.2.3 Advanced Air Management

Throttling, Variable Geometry Turbocharging (VGT), EGR (either internal, high pressure, or low pressure), and a range of variable valve timing technologies (VVT, from simple mechanical systems to fully-flexible electro-hydraulic systems) are mature technologies with significant market penetration. The combination of these technologies allows a precise control of the cylinder charge and its concentration and, for the most advanced VVT concepts, the possibility of controlling it for every specific engine cycle and even implementing a simple way of cylinder deactivation for partial load operation [20].

Major challenges are still attached to the coordination of the systems, as cross-interactions exist between the most common feedback control variables and any tracking error usually results in significant emission peaks or the inability to supply the requested torque. Model-based estimation and control methods of different complexity are being used with success [21], and most production controllers use them to some extent – usually in combination with low level PIDs.

4.2.2.4 After-Treatment

The proper after-treatment operation is mandatory for fulfilling current and future emission standards. State-of-the-art engines combine a list of different bricks specifically targeting a number of functions. In the case of stoichiometric combustion, three-way catalyst (TWC) allows an all-in-one solution for HC, NOx, and CO abatement, which only needs to be combined with a particulate filter for trapping soot produced in the combustion. Lean combustion needs dedicated oxidation (for CO and HCs) and reduction catalyst, plus the particulate filtering. It is possible to combine several functions for example in impregnated particulate filters that may act as SCR too, or oxidation catalysts able to trap NOx during the cold operation of the system, as depicted in Figure 4.3.

A set of actuators exist, as dedicated injection settings for either promoting the emission of HC that can be burnt in the exhaust line (targeting particulate filter regeneration) or with high exhaust temperature (for fast warming of the ATS), the oxygen excess in the exhaust, to the injection of an aqueous solution of urea acting as a reductant agent for the SCR. After-treatment control is a complex problem since many of the components need a tight control of the operation temperature and concentrations of the exhaust species, which is not always possible due to changing driving conditions, the need of regenerating the particulate filters, and the start well below the catalysts light-off temperature. Next-generation engines are expected to include electrical heating for accelerating the ATS light-off.

Sensing technologies are well developed, and ubiquitous lambda sensors may be complemented with temperature, pressure, and NOx concentration sensors.

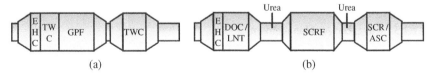

(a)   (b)

Figure 4.3 Aftertreatment configuration for next-generation gasoline (a) and diesel (b) engines. EHC: electrically heated catalyst; TWC: three way catalyst; GPC: gasoline particulate filter; DOC: diesel oxidation catalyst; LNT: lean NOx trap; SCR: selective reduction catalyst; SCRF: particulate filter with SCR function; ASC: ammonia slip catalyst.

Particulate matter and NH_3 sensors also exists but with a lower penetration in the market. From a control perspective, a number of problems have been identified, depending on the combustion mode and the ATS architecture: closed-loop fueling for TWC operation optimization [22], operation of TWC in lean-burnt combustion [23], optimal dosing of the reductant agent in SCR [24], solving cross-sensitivity of NOx sensor to NH_3 slip [25, 26], ATS ageing observation [27, 28], fast warming of ATS [29], control, avoidance of unexpected regeneration events [30], among others.

4.2.2.5 System Integration and Calibration

System integration and the optimization of the engine operation (including fueling, combustion, charge control, thermal management, and after-treatment operation) is still the major challenge of the modern ICE. There has been a continuous increase in the degrees of freedom of the engine attached to the progressive deployment of new technologies and the penetration of electronic control on different subsystems. Some of the actuators, such as injectors, ignition, and fully-flexible VVT, have the possibility of actuating several times along the same engine cycle, and their control references may be adapted cycle-to-cycle and cylinder-to-cylinder. In addition, the engine must adapt to significant variations of ambient and use conditions, since the same engine may be used in hot to cold weather, sea level, or high altitude, and with different payload and driving style. Finally, several operation modes are also possible (e.g. after-treatment regeneration mode, low combustion noise mode, fast catalyst warming mode, etc.), and fuel quality can also vary.

Current electronic control units (ECU) have state machines shifting between control algorithms, and each one of them has a number of parameters – usually scheduled in lookup tables – that allow tuning the final operation of the control system. The number of parameters that may be varied through the so-called *calibration* process is more than several tens of thousands. As it may be expected, finding a suitable (not yet optimal) combination of those parameters is a major challenge. With the onset of RDE procedures, were the vehicle is tested in real driving conditions with no *a priori* knowledge of the test cycle, the calibration of the control system has become a problem of major relevance.

Model-based control has been used as a way of reducing the number of tests needed for the calibration process [31]. Modern ECUs embed a growing number of physical models – usually referred to as Mean Value Engine Models (MVEM) – and a number of model-based estimations – known as virtual sensors – which are used for feedback control or for diagnostics, no cohesive approach has been developed to date. While a range of works have covered the use of optimal control for different engine systems with success [21, 32], industry adoption of these techniques has been limited because of some industrialization requirements, e.g. the capability of tuning the control system once in production. As it may be expected, the calibration process needs a significant human intervention and intense testing in both laboratory and real driving conditions. Digital twins have been suggested as a way of reducing the test intensity, and a range of experiment design and automated calibration tools have been developed, including the application of different machine learning approaches as reviewed in Garg et al. [33].

4.2.2.6 Connectivity

Powertrain control and diagnostics can benefit from the systematic fleet-wide recording of data along the vehicle service life. Real-life data provides an extensive assessment of the control system performance, and makes possible detecting challenging situations in need of improvement. Compensation of sensor and engine bias, by means of statistical approaches, may be done, although this would need deploying calibrated sensors during technical inspections in a subset of the fleet [34]. Over-the-air ECU software and calibration update is already a possibility, and control and diagnostics software can be updated along vehicle service life. In this sense, it has been proposed to seed the fleet with different calibrations or control system versions, and implement distributed learning approaches [34]. However, the implementation of unsupervised adaptive control techniques is still far from market implementation: on one hand, engine control involves several sensitive systems that could result in safety risks; on the other hand, several regulating bodies specifically ban such techniques, as they have been used in the past for circumventing the emission regulations.

Nonetheless, it may be expected that the regulatory framework will soon be adapted to the connected vehicle paradigm, softening some of the current restrictions in order to benefit from some of its advantages. In this sense, OBM implementation, together with connectivity, may provide the regulatory bodies a sufficient evidence of the fleet compliance with the emission regulations, and may allow emission geofencing policies [35] (e.g. PHEV only allowed to run the ICE out of highly polluted areas). In OBM implementation, the measurements provided by the on-board sensors will be sent over-the-air in order to perform various activities, in which a nonexhaustive list is provided below:

1. Real-world emissions performance: The different sensors installed in the vehicle will be able to transmit the exhaust after-treatment efficiency of the vehicles over their lifetime by supervising the tailpipe levels.
2. Detect high emitters: By measuring the quantity of pollutant emitted over a defined distance, every vehicle in the fleet may be checked against the regulation limits.
3. Data streaming: In its final implementation, the stream of data is not expected to only go from the vehicle to the final user, by could also work the other way

around. OEMs, policy makers, or organizations might request emissions monitoring from specific driving conditions or vehicles. Blockchain techniques have been suggested as a way of securing the information.
4. Statistical fleet analysis: The amount of data generated will also contribute to future transportation analysis and could therefore contribute greatly to improving this sector (e.g. average trip, driving patterns, consumption, and emissions).

Diagnostics is another field that is expected to benefit from connectivity, and the use of artificial intelligence is an open research field for harnessing the huge data stream. In opposition to centralized cloud architectures, where all data is uploaded to a central server for its processing, decentralized architectures as in Figure 4.4 will allow deploying the intelligence on the edge (i.e. vehicle) and restrict the data to be streamed. As opposed to classical OBD algorithms, next-generation diagnostic systems will be able to learn from the fleet data, find correlation between vehicles, and use historical records for analyzing the performance of the system along its life, both at component, system, vehicle, and fleet levels [34].

Other advantages of the connected vehicle beyond powertrain control and diagnostics will be presented in Section 4.5.

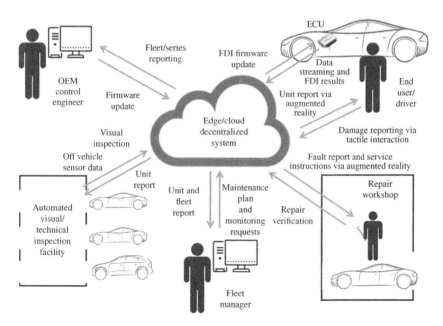

Figure 4.4 Decentralized connected diagnostics framework, as proposed in ASSIST-IoT. Source: [34]/Elsevier.

4.3 Electrification

4.3.1 Electric Motors and Drives

The requirements for electric motors and drives depend on the EV or HEV powertrain configuration. Several hybrid architectures (with electric machines on the order of 100–200 kW) can be commonly found in production systems. In a series hybrid, the first electric machine is mechanically coupled directly to the combustion engine and electrically coupled to a battery and a secondary electric drive motor. This architecture allows the engine to always operate at the peak efficiency for any given power level, but the main challenge is the low overall efficiency due to multiple energy conversions. In this configuration, the engine is typically used as a range extender. Parallel hybrids have the potential for higher overall fuel efficiency because the extra energy conversions can be avoided under some operating conditions. Parallel hybrids typically utilize a speed-coupled ICE engine and electric machine resulting in fewer degrees of freedom. In this configuration, the fast-acting electric machine torque is regulated to achieve the desired energy balance and fill in for the ICE during transients. A combination of these two HEV architectures can be realized, with the promise of greater fuel efficiency, by including clutches, additional electric machines, and planetary gears, which increases the complexity and control challenges. Since the cost of improving fuel economy is a primary consideration, 48V micro and mild hybrids are becoming more popular as they have the highest fuel economy improvement per dollar. For these systems, the size and location of the electric machine along the drive train are commonly indicated by the P0-P4 motor configurations. The P0 configuration, also known as a belted starter-generator, replaces the function, and position of a typical alternator connected to the front-end accessory drive. This configuration enables engine start–stop, which gives the highest fuel economy benefits per dollar, but the torque and hence power level is limited by belt slip to between 5 and 15 kW [36, 37]. As the motor moves backward into the transmission and drive axle as shown in Figure 4.5, larger motors are required, but the vehicle can sustain pure electric operating mode under a wider range of speeds and accelerations. Finally, high-speed electric motors are well suited to provide electric boosting, either in the form of an electric turbocharger or super-charger, and can further extend the operating power of the internal combustion engine enabling downsizing of the combustion engine to boost efficiency without sacrificing peak power. In many cases, e-boosting must be combined with a P0 machine in order to re-charge the 48 V battery, and a major control challenge is determining whether to provide torque assist or electric boosting given the system level constants including real-time battery power capability and switching time for the clutch [38].

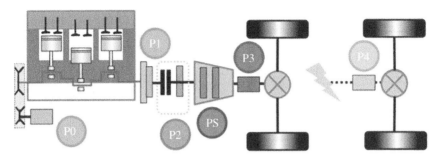

Figure 4.5 Potential locations for 48V motors within a Hybrid electric vehicle Powertrain, Starting from the front of the engine with the smallest motor at the P0 location, motor power is typically increasing moving backward toward the rear drive axle. The addition of clutches in the P2 and higher enables all-electric driving with the engine off. Image source (BorgWarner).

Three-phase AC induction motors (ACIM) and Permanent Magnet Synchronous Motors (PMSM) are generally used as traction motors in electric vehicles. Instead, higher speed capable and low-cost Switch Reluctance Machines (SRM) are used in E-boosting applications [40]. Tesla gained notoriety for using ACIMs in the first generation roadster and Model S. For ACIMs, torque is generated in the motor by producing a sinusoidal voltage with an asynchronous frequency and the torque produced is proportional to the slip or frequency difference. When the frequency of the AC voltage is the same as the motor, no torque is generated. Permanent Magnet motors are the more popular choice for EVs due to their higher efficiency and power density, but the control is more complex. Control of PMSM is typically implemented in the transformed synchronous rotating reference frame using the measured position of the motor from an encoder or resolver [41], see Figure 4.6.

For control of the electric motor in EVs, vibration, and auditory noise are particularly noticeable by the driver due to the quieter vehicle operation. Therefore, many control strategies focus on reducing Noise Vibration and Harshness (NVH) throughout the drivetrain. For PM motors, there are significant opportunities for coupled control and diagnostics of the motor and inverter because currents can be injected into the motor, which does not generate any torque [42]. Faults that can be detected range from inter-turn winding shorts, coolant system failures, bearing failures, and demagnetization [43].

4.3.2 Lithium-Ion Batteries

With the introduction of Lithium-ion batteries into automotive powertrains several new control challenges have arisen over the past two decades in HEV and battery electric vehicles (BEV). For passenger car HEVs, the battery size is

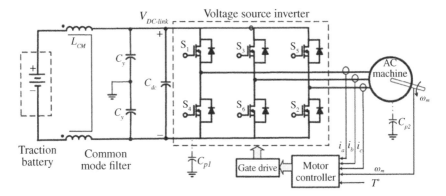

Figure 4.6 Voltage source inverters are typically used to drive three-phase electric machines. Industry trends toward the adoption of wide band-gap semiconductors enable operation at higher voltages and switching speeds. The improved efficiency (lower conduction and switching losses) also reduces the inverter cooling requirement. Control of Permanent magnet electric machines requires estimation of the flux linkages within the machine and hence relies on motor position sensors to achieve high-performance torque regulation. Source: [39]/IEEE.

typically small, between 1 and 8 kWh, and capable of providing less than 20 miles of all-electric driving range in systems with the largest batteries. Depending on the hybrid architecture, the voltage is typically between 48 V and 160 V DC, and the power requirement can be as high as 100 kW. Therefore the stresses on the battery for HEVs are typically much higher than BEVs since the goal is to shift the operating point of the internal combustion engine (ICE) to a more efficient operating point with the lowest overall system cost. Since the battery cost is significant, HEV systems aim to use the smallest economical battery. The battery power and current are typically normalized by the battery capacity in (kWh) or (Ah) in order to make comparisons across packs or cells of different sizes. A 1°C-rate current would deplete the battery in one hour. For HEVs the peak currents could be as high as 25°C-rate, which increases the need for cooling to combat resistive losses in the battery, which increase with the square of the current.

On the other hand, BEVs typically have battery systems capable of providing between 200 and 300 miles of electric range. If we consider an average energy consumption of 285 Wh/mile, this requires pack capacities between 60 and 100 kWh. For a 200 kW electric motor, the peak battery power needed to propel the vehicle up to highway speeds (for a duration of less than 10 seconds) is only 2 or 3°C rate for a typical BEV, which does not present significant challenges for active cooling of the battery during normal driving, where peak acceleration events are short duration and relatively low duty cycle. However, consumer demand for faster charging times drives the design requirement of the battery cooling system. Present EV systems can achieve sustained charging of rates 120 kW, or 7 miles of range per minute of charging. The US DOE, has identified a target of 15-minute charging capability to bring the battery from 10% to 90% SOC to be competitive with conventional engines running on liquid fuels. The 15-minute target corresponds to an average C-rate of 3.2. For a 100 kWh BEV, it is anticipated that the charger should be capable of at least 400 kW to meet this average charging rate. To achieve these current rates the cooling system should be oversize compared to the requirements for propulsion [44]. Furthermore, advanced diagnostic algorithms are needed to ensure the charging system, cooling system, and battery are operating properly to ensure safety during DCFC. Overcharging or overheating of lithium-ion batteries can cause catastrophic failure, often called thermal runaway, resulting in a combustion that is virtually impossible to extinguish [45–47].

The primary control challenges related to integrating batteries into the powertrain can be grouped into three main categories: Energy management which relies on accurate SOC estimation, enforcing the protection and power limits commonly referred to as state of power (SOP) estimation, and performing Diagnostics and Prognostics commonly referred to as state of health (SOH) estimation. The two main types of battery models that are commonly used for these purposes include equivalent circuit and physics-based models. Equivalent circuits are most often used in industry due to their reduced complexity and straightforward parametrization [48]. These models utilize parameters (resistances and capacitance) that vary with temperature, SOC, and battery health. Since these models rely on interpolation or lookup tables, they tend to perform well over the operating regions for which experimental data was collected to parametrize the model but suffer from poor performance when extrapolating. Therefore many experiments must be performed to parametrize the model. There is also some debate as to whether accelerated aging tests accurately represent the expected battery degradation corresponding to real-world usage patterns over the life of the cell which should be more than 10 years. These parameters could also be learned or identified online during the operation of the vehicle [49–52]. Electrochemical models are often based on porous electrode theory [53, 54]. These models include sets of nonlinearly coupled partial differential equations corresponding to the diffusion of lithium within the electrolyte and solid phases of the battery. Reduced order electrochemical models such as the electrolyte enhanced single particle model (SPM-E) have also been successfully used for BMS applications [54–56]. Physics-based models are preferred when considering the effects of temperature on performance and predicting capacity loss due to various aging mechanisms. Over time and usage Lithium-ion batteries degrade, and therefore have a finite usable life in automotive applications. The battery end-of-life is typically defined as the point at which the pack cannot deliver the required power or driving range. Several factors, including the depth of discharge and C-rate when cycling and the time spent stored (calendering) at various temperatures and SOC conditions contribute to capacity loss and an increase in the internal resistance of the cell. Storing the battery at an elevated temperature (above 45°C) and high SOC increases the rate of aging. For passenger car applications, the primary degradation is calendar aging since the duty cycles are typically less than 10% or one cycle per day. Physics-based models are useful for mitigating the deleterious effects of fast charging, by placing constraints on the electrochemical states to avoid over-potentials in the negative electrode that would cause lithium plating [57].

The battery is typically modeled in Galvanostatic operating mode, with current as an input, and voltage as an output. The simplest form of a battery equivalent circuit model, which can be used for SOC estimation, includes a single dynamics for the SOC corresponding to a pure integrator,

$$\dot{z} = -\frac{I}{Q} \tag{4.1}$$

where z is the SOC, I is the input current and Q is the capacity in (A s). The output voltage $V_T = V_{ocv}(z) - IR$ can be modeled by the cell open circuit voltage which is a function of the SOC and an ohmic resistance loss. The problem of SOC estimation is mostly solved using an Extended Kalman Filter (EKF) due to the nonlinear relationship in the $V_{ocv}(z)$ function. The problem of SOC estimation is relatively easy for higher voltage cathode chemistry such as Nickle-Manganese-Cobalt (NMC) that has more than 1 V of change across the operating range. However, iron-phosphate (LFP) cathodes, which are recently gaining popularity due to the avoidance of cobalt, have a flat voltage profile and hysteresis in the measured terminal voltage and therefore are more difficult to accurately gauge SOC. The SOP estimation is closely coupled with the thermal management of the battery system as the internal resistance is a strong function of temperature, increasing sharply at low temperatures. Power capability can be readily inferred from either the ECM or SPM models via model inversion, and the limits are typically set based on a 10s time horizon.

The primary battery warm-up strategy used in cold temperatures is preconditioning when charging (or plugged into a power source) to mitigate the impact of low ambient temperatures. However, during operation in BEVs, there is little waste heat to provide for occupant comfort or to maintain optimal battery temperature, and therefore battery energy is used, further depleting the vehicle range. It is not uncommon in cold climates for BEVs to have a 50% reduction in achievable range compared to a moderate climate. Higher efficiency heat pumps can be used to improve the system performance but also increase the system cost.

Diagnostics and prognostics for lithium-ion batteries rely on accurate SOH estimation. Battery internal resistance is easy to characterize from usage data, but the capacity can be much more difficult to determine without a deep discharge and full charge cycle that could rarely occur during normal usage [58].

Therefore physics-based models of the battery can provide greater insights into the degradation states (internal to the battery). In particular, utilizing modeled information about the individual electrode half-cell potentials can be used to attribute the capacity loss to each of the individual electrodes [59]. This is critical for predicting the future rates of capacity loss for any given usage profile.

4.3.3 Outlook and Future Challenges

It is anticipated that connectivity and vehicle networking will change the landscape of transportation. In addition, BEV's have the potential to shape our future electrical grid and power systems. Congested and aging power transmission infrastructure will need to be upgraded to accommodate large-scale EV adoption, DC fast charging, and penetration of distributed renewable electric power generation [60]. Not only are vehicles mobile power sources that could transport and store energy, but they could provide significant relief to the power transmission network if bi-directional charging becomes standardized (either in front or behind the meter with Vehicle to Grid V2G, and Vehicle to Building V2B capabilities). Outside of the vehicle, battery energy storage systems (BESS) can be co-located with DC fast chargers to reduce the impulse loads on the grid. Further information from vehicle connectivity could also be used for energy management of the infrastructure to anticipate future load demand and pre-charge of BESS. Future research is needed to address these large-scale energy optimization problems which could include both the transportation network and the electric power grid. The value of a lithium-ion battery pack at the end of its life in a vehicle is still uncertain, in the near term most batteries will be scrapped and recycled. However rapid and accurate diagnostics and prognostic algorithms could enable batteries with significant remaining useful life to be repurposed for second life applications in stationary energy storage before eventually being recycled. The requirement for a "Battery Passport" in Europe to help determine what information should be tracked for the valuation of batteries is expected to help shape and enable the industry for lithium-ion battery recycling. A key enabler for second life would be sharing data about the pack SOH and usage conditions during its first life in the vehicle.

4.3.4 Fuel Cells

Low-temperature Proton Exchange Membrane (PEM) fuel cells are most relevant for light-duty automotive and heavy-duty applications. A PEM fuel cell converts hydrogen fuel and oxygen from the air, into electrical energy over a platinum catalyst. The membrane (typically made of Nafion) is a proton conductor and electrical insulator which keeps the hydrogen and oxygen gas separate. Hydrogen is readily oxidized at the Pt surface and protons travel through the membrane. The electrons are forced to go through an external circuit to reduce the oxygen combined with the proton to form water. The voltage produced by this reaction is relatively low when producing power, around 0.65 V per cell. Therefore many hundreds of cells are connected electrically in series to reach system voltages of 400–800 V. The only byproducts of this reaction are heat and water. The fuel cell operates at relatively low temperatures between 60 and 80°C, and therefore the liquid cooling system requires a larger radiator than a comparable-sized internal combustion engine, and the thermal management is more complex as the stack operating temperature can significantly impact the system efficiency. On the high-temperature side, the thermal management controller must place stringent constraints to prevent damage to the polymer membrane [61]. The fuel cell stack voltage decreases with the amount of current drawn from the stack, which is referred to as the polarization curve. The auxiliary power consumed by the cooling pumps, fans, and the air supply system increases with the square of the fuel cell current, and therefore a peak operating efficiency is achieved near 25% of the system's peak power. The set of subsystems required to operate the stack are collectively referred to as the balance of plant or BOP as shown in Figure 4.7.

The thermal and water management problems are intrinsically coupled in PEM systems as the membrane needs a good amount of water to achieve high proton conductivity (lowering the restive losses that generate heat), but due to the required operation below 100°C can have two phases of water present in the stack. Excess water that cannot be carried away in the vapor phase with excess air supplied to the cathode, and condensing and liquid water building up in the catalyst layer or gas channels impede the flow of oxygen to the catalyst sites which drastically reduces FC efficiency and limits the power capability. The membrane also absorbs water to be in equilibrium with the relative humidity of the gasses in the anode and cathode gas volumes, and hence thermal cycling can also drive changes in the membrane water content. The membrane swells with absorbed water so this mechanical cycling leads to fatigue, pinholes, and failure of the membrane [62, 63]. This is one source of degradation that needs to be addressed through careful control of the system, including hybridization and power split since this impacts the thermal loading and sizing of the system. Corrosion of the carbon which supports the Pt catalyst is another major concern for durability, which requires precise control of the airflow as oxygen starvation due to insufficient flow is damaging and excess flow lowers the system efficiency increasing fuel consumption [64]. Carbon corrosion is also an issue during the turn-down and start-up of the system [65]. To alleviate some of these challenges fuel cells are typically hybridized with a lithium-ion battery pack or supercapacitor and buffered with a DC/DC converter to regulate the power drawn from the fuel cells stack. Therefore, in many modeling and control formulations, we can assume the stack current is a controllable input to the system.

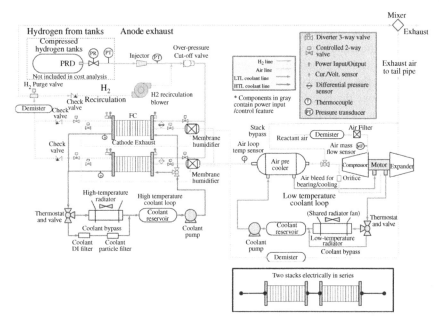

Figure 4.7 Heavy-Duty Fuel Cell Systems include multiple fuel cell stacks, high and low-temperature coolant loops for thermal management, a hydrogen recirculation loop to maximize hydrogen use, and a compressor/expander to provide the required airflow. All of these interconnected subsystems require precise control to achieve robust system operation and avoid damage to the stack. In addition to the battery and motor drive components typically found in an electric vehicle, a FCEV would also require an additional DC/DC converter to regulate the power split between the battery and fuel cell stack. Source: [66]/U.S. Department of Energy.

4.3.5 Outlook and Future Challenges

In the United States, limited hydrogen fueling infrastructure and rapidly decreasing costs for Lithium-ion batteries have given BEV's an advantage, and they will likely displace FCEV in passenger car applications. Battery energy density is not high enough for heavy-duty, long-haul trucking, marine, and rail applications. Therefore, PEMFCs remain a viable alternative to combustion engines in these vehicles. The US Department of Energy has set ultimate targets for fuel cell systems in light-duty automotive applications (corresponding to an 80 kW net stack) at US$30 per kW net or roughly 1/2 of the current system cost. For heavy-duty applications with a 275 kW stack size, the durability requirements are much more stringent due to the higher usage duty cycles and this is reflected in the cost target of US$60/kW net, which is roughly one-third of the 2022 cost estimates [66]. The areas in which the department of energy is prioritizing funding for fuel cell system development are cost and durability. The air management system makes up the largest cost among the balance of plant components shown in Figure 4.7 and has the largest potential for cost savings. However, like a boosted and down-sized ICE, the transient performance of the device is limited by how quickly the air supply can be ramped up and remains an area of active research and development. At the vehicle systems level, coupled thermal and energy management strategies have the largest potential to improve system durability. In early pilot programs for fuel cell buses, the majority of downtime was caused by BOP system component failures, such as the air compressor. Improved diagnostics algorithms for PEMFC are needed to enable the prognostics and predictive maintenance for fleet management that will further lower the cost of operating these systems.

4.4 Driver Assistance Systems and Automated Driving

Over recent years the major developments in Advanced Driver Assistance Systems (ADAS) have revolved around increasing the level of autonomy in the vehicle aiming to eventually reach full autonomous driving. Often, the degree of autonomy of the vehicle is classified according to the SAE Autonomy Level scale [67]. As shown in Figure 4.8, starting from completely manual vehicles (level-0), the degree of automation increases to single function automation (level-1) and to multiple functions automation (level-2). Further increasing the automation, at level-3 the vehicle has full control without constant driver supervision, but the driver must be able to be alerted and called to intervention, if the system needs. Such interventions become unnecessary at level-4, under the restriction to operate on specific conditions and areas, e.g. by geofencing, while this last restriction is finally removed at level-5 to achieve full autonomous driving, including the case where no physical controls are available for the driver.

At the date this chapter is written, level-1 and level-2 are the quite common in production vehicles, where functions such as adaptive cruise control, lane keeping, automated parking and traffic jam assist are available, and often integrated. In fact, level-1 and level-2 functionalities have been developed for several years as part of ADAS packages, where control played a significant part in the development [68–70]. Currently, the major development focus for future production vehicles is on achieving effective and reliable level-3 functionalities, where as few interventions as possible are requested, and the driver is requested to intervene early enough to be able to assess and handle even complex scenarios. Such systems are relatively close to being in full vehicle production lines, and are expected to be fairly common by 2030, at least under certain conditions.

Instead, the current automated driving research focuses on level-4 and level-5 developments, to achieve automated driving in special and general conditions, respectively. Here the challenge is that the Autonomous Driving System (ADS)

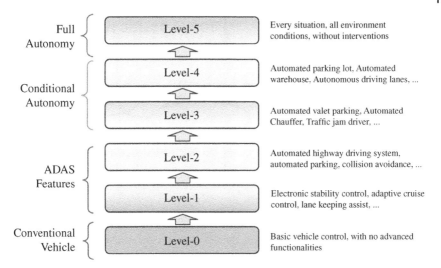

Figure 4.8 Autonomy levels according to SAE standard, and related vehicle features.

must be able to autonomously handle any situation, including unexpected and faulty behaviors [71]. Even though at level-4 this may be restricted to a specific area, where also infrastructure support may be available, considering and properly handling all events that may happen on a general road is still quite challenging. Because of that, recently several organizations started to investigate ADS for special operations and for closed areas, where the behaviors of the environment are more restricted and hence fewer scenarios need to be accounted for. Some examples include fully autonomous lanes/roads, factory areas, and logistics centers [72, 73], and more are arising.

While AD comprises multiple functionalities, such as positioning, sensing, and perception, knowledge representation, the planning and control features will have, contrary to many other applications [74], high visibility, since it will be what the driver/passenger experiences as the physical result of the ADS. Thus, control systems bear significant responsibilities, and at the same time opportunities, in the development of ADS. Indeed, the road toward level-5 ADS deployment is still long, but the control research aimed at ADS will have opportunities to be included in ADAS packages deployed much sooner than ADS ones, as it is already happening.

4.4.1 ADAS/ADS Architectures

The ADS will keep evolving from lower autonomy levels, consistent with ADAS, to higher ones over many years and product cycles. In order to simplify code re-use during such incremental development, while retaining flexibility and

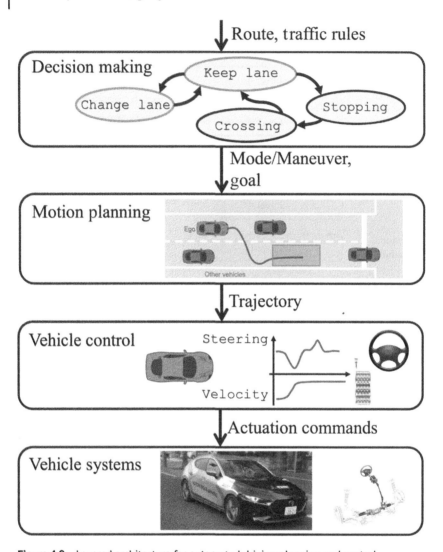

Figure 4.9 Layered architecture for automated driving planning and control.

performance, it is natural to imagine that evolving versions of ADS planning and control features will be organized in modules. An example of modular architecture for planning and control, specifically, layer-based, is shown in Figure 4.9, where from top to bottom, the *decision-making* layer determines the next behavior/action that the vehicle needs to execute, e.g. changing lane, stopping at an intersection, crossing such intersection, and consequently, the next goal for the motion planner. The *motion planning* layer generates a trajectory to achieve such goal, while satisfying traffic rules and avoiding collisions with other actors on the road, e.g. cars, bicycles, pedestrians, and the *vehicle control* layer tracks such trajectory by commanding the vehicle actuators, steering, accelerator, brake, and, in some cases, gearbox.

Modular software architecture allows for flexibility in updating specific modules or introducing new modules, without changing the others, and hence allows expandability and life-cycle maintainability of the ADS. Also, it enables usage of different algorithms, at different execution rates, and with different abstractions of the process to be control, i.e. the vehicle behavior and the surrounding environment, and hence with different computational load. Finally, from a product development process perspective, it enables shared development based on interface specification without the need to expose the full implementation, for privacy or IP reasons. This is particularly important when multiple different companies work on different ADS modules, which is common practice in automotive development between OEMs and Tier-N suppliers. In fact, while monolithic architecture, so called end-to-end, may eventually be considered when ADS is mature enough, it seems that most ADS projects starting from the DARPA Challenges have opted for such modular architectures [75–77], to keep growing their capabilities by expanding the modules and/or the module features without requiring a full re-design.

However, modular architectures also present challenges. While the single modules may exhibit the desired behaviors when tested in a stand-alone setup, they may behave differently from what is desired when they are integrated. For instance, the decision-making layer in Figure 4.9 may determine an action that the motion planner may not be able to achieve, and thus this may fail or waste a large amount of computations – and hence resources and time – to compute an impossible trajectory, or the vehicle control may not be able to track precisely the trajectory generated by the motion planner. These issues arise due to multiple factors, including the different rates at which different layers are operating, different abstractions on the prediction models that each internally use, and different computational resources available. However, several methodologies developed in control research can be used to support the integration process, from reducing the issues, to ensuring formally that integration issues do not occur. In fact, control-loop composition is a classical problem in control design, for which several tools are available, and several recent methods, such as reachability, invariance, and Lyapunov methods [78–80] can be used to ensure proper operation when modules/layers are integrated. Some contribution in this area have started to appear see, e.g. [81], yet more work will be required over the next few years.

A particularly important challenge will be on integrating the control stack with the perception stack. While both are extremely important, and strictly critical, functions of the ADS, very often they work using different data and models and their integration is merely based on information passing. However, this ignore the intrinsic feedback loop between perception and control, where indeed perception quality affects control decisions, e.g. the need to avoid areas for which you have only limited available information, and control decision affects perception quality, e.g. the vehicle trajectory affect the sensors' unobstructed field of view and the areas of the road where these are pointing. The integration of perception and control in ADS is still at very early stages, but can build on research in visual servoing [82], vision-based control [83], SLAM [84], and perception aware control [85], which have reported several results, at various levels of maturity.

4.4.2 Current and Future ADAS/ADS Features

Several ADAS/ADS features have been or will be introduced, requiring different control technologies to support them in the different layers of the architecture shown in Figure 4.9. Among the ADAS features that are already quite common in the market we can find Adaptive Cruise Control (ACC), Electronic Stability Control (ESC), Lane Keeping Assist (LKA), and Automated Emergency Braking (AEB). Most of these features requires operation of the vehicle control layer, and fairly standard control technologies, such as PID, loop-shaping or simple state-space control. However, benefits in terms of performance and especially operating envelope, i.e. when the system can function successfully, can be achieved using more advanced control methods, especially robust control and model predictive control (MPC) [86–89]. For instance, the recovery capabilities of ESC can be increased by MPC [90] and performance guarantees of ACC or AEB can be obtained using control barrier functions [91]. In fact, most of these technologies are now mature for mass deployment and are starting to appear in production vehicles [81]. Thus, from a technical perspective, several current and future technologies for implementing a stand-alone vehicle control layer are well established, and most of the technologies development at this layer may focus on proper integration with additional layers for more advanced functionalities [92, 93].

Another set of features that is starting to become common are those related to level-2 technologies that integrate control of multiple subsystems to enable "hands-off" driving, where the driver relinquishes control of the entire vehicle, yet must maintain constant supervision of the system. These include the Automated Highway Driving System (AHDS) and Automated Parking (AP). In terms of technologies, beyond the vehicle control layer these features require also some functions at the planning level, more and more powerful depending on the operating envelope, i.e. scenarios where the system is enabled to operate. Current planning technologies are quite mature for these features, in several – if not all – the desired scenarios. When the same features will be extended to level-3 system, where the driver is no longer in constant supervision but may need to be called back, the key challenge will be providing enough time for the transition from automated to manual operation in such intervention situations. This will require appropriate look-ahead of the planning and control system, thus certainly requiring some form of predictive control and longer horizon planning, as well as appropriate technologies for informing the driver of the current environment, the scene awareness, and smoothly transitioning control from the ADS to the manual system, the transition hands-off. The research on these technologies, as well as the research on computationally-efficient planning and control algorithms, which will be needed to lengthen the prediction horizon within the limited on-board computing capabilities in the vehicle, has been very active in recent years.

When moving to the ultimate feature of full autonomy in limited and unlimited environments, level-4 and level-5, respectively, a key technology will be the determination of the operating mode of the vehicle: whether crossing an intersection or changing lane, whether merging into traffic or remaining on the merging lane. These decisions are necessary for determining an appropriate target for the motion planner and are often carried out by a decision-making module, the top layer in Figure 4.9. Currently, these are some of the most active areas of research in planning and control of automated driving, where research tools in control theory, such as reachability methods, invariant sets, Lyapunov methods are being investigated [78–80], as well as integration of those with techniques from machine learning [94–96]. A differentiating factor in level-4 and level-5 features will be the information about the environment. Since level-4 features operate in a limited-area (geofencing) precise mapping information will be expected, and likely support from infrastructure in terms of information and possibly control sharing. Thus, for level-4 a key technology will be the integration of control and communication, which has been very active in the last 15 years, including possibly crowd-sourcing information sharing, as well as cloud-computing enabled control [97–100]. For level-5 such assumptions may not be possible, and hence a major focus will be on robust control, including fault-tolerant control, intended in the sense that when some of the predictions on the environment behavior will turn out to be incorrect, the ADS must still ensure safety, possibly at the price of reduced performance/comfort, e.g. hard brake to avoid a vehicle cutting in in front of ours. Finally, since the ADS may not be able to rely on infrastructure support, robust sensing and perception will be necessary and to this end a robust integration of control with perception, including exploiting the synergy and relations between control and perception quality [101, 102], see Figure 4.10, will provide significant benefits.

4.4.3 Deployment of ADAS/ADS Features

Besides the research to develop and apply control methods for new ADAS/ADS features, the deployment into mass production vehicles requires significant

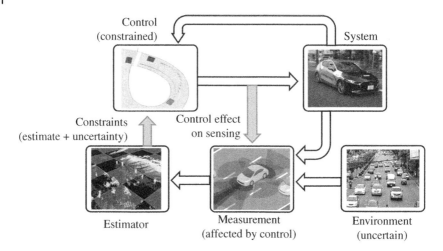

Figure 4.10 Feedback interactions between control and perception of the environment arising in operation of a fully autonomous vehicle in a dynamically changing environment.

efforts in terms of validation, calibration, and productization. Figure 4.11 shows exemplary stages of a deployment process, from research to product. As opposed to classical control laws used in automotive in the past, which are based on evaluating a function, the control approaches being investigate for future ADAS/ADS features are often based on algorithms, e.g. MPC, sampling based planning, reinforcement learning. This amounts to larger program codes that are to be written in a way that is compatible with the safety standards required in automotive applications, e.g. MISRA-C [103], and validated according to such standards. To

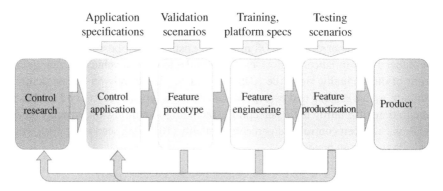

Figure 4.11 Schematic of deployment process steps from control research to product, information added at each time, and update loops.

this end the latest technologies in safety-by-design, automated verification and validation, and model-based design and validation may significantly contribute to the timely deployment of advanced control technologies for new ADS/ADS features. All these technologies are somewhat familiar for control researchers, but they will need to be taught to practitioners in the engineering development centers, and to calibration specialist responsible for in vehicle analysis and testing. Thus, a significant effort should be aimed at developing training resources and classes to disseminate such knowledge to practitioners that are not control system researchers and, in the largest majority of cases, do not hold a PhD in control. At the same time, the development of tools that automate part of the deployment and testing, such as automated calibration methods, tools for analysis from experimental data, models, and digital twins of vehicle operations, may significantly speed-up the deployment.

As one example, the recently developed ADAS-ECU from Mitsubishi Electric [81] includes ADAS functionalities handled by advanced MPC algorithms. Besides the research efforts in develop the control algorithms and the numerical solvers, an equal if not larger effort has been in training engineers in the development center through a series of whiteboard and hands-on classes, refactoring the algorithm codes to be compliant with safety standards, evaluating the algorithms in safe conditions using model-based validation and hardware-in-the-loop platforms, and automating parts of the calibration process for the control algorithms and the numerical solver. Both the research and the process development efforts are necessary to reach the stage of deployment for new features based on advanced control methods.

4.5 Connected and Integrated Transportation Systems

Connected and automated vehicles (CAVs) (Figure 4.12) have attracted considerable attention over the last decade since they provide the most intriguing opportunity for enabling users not only to improve powertrain performance but also to better monitor transportation network conditions and make better decisions for improving safety and transportation efficiency [104, 105]. In a transportation network with CAVs, we can consider the vehicle as part of a larger system, which can be optimized at an even larger scale. Such large-scale optimization requires the acquisition and processing of additional information from the driver and conditions outside the vehicle itself. This requires the addition of new sensors and/or better utilization of information generated by existing sensors. The processing of such multi-scale information provides new opportunities for developing signifi-

Figure 4.12 Connected and automated vehicles in a traffic environment. Source: zapp2photo/Adobe Stock.

cantly new approaches in order to overcome the curse of dimensionality. It seems clear that the availability of this information has the potential to reduce traffic accidents and ease congestion by enabling vehicles to more rapidly account for changes in their mutual environment that cannot be predicted by deterministic models. Likewise, communication with traffic structures, nearby buildings, and traffic lights should allow for individual vehicle control systems to account for unpredictable changes in local infrastructure.

Recognition of the necessity for connecting vehicles to their surroundings has gained momentum. Many stakeholders intuitively see the benefits of multi-scale vehicle control systems and have started to develop business cases for their respective domains, including the automotive and insurance industries, government, and service providers. The availability of vehicle-to-vehicle (V2V) and vehicle-to-infrastructure (V2I) communication has the potential to ease congestion and improve safety by enabling vehicles to respond rapidly to changes in their mutual environment. Vehicle automation technologies can aim at developing robust vehicle control systems that can quickly respond to dynamic traffic operating conditions. With the advent of emerging information and communication technologies, we are witnessing a massive increase in the integration of our energy, transportation, and cyber networks. These advances, coupled with human factors, are giving rise to a new level of complexity [106] in transportation networks. As we move to increasingly complex emerging transportation systems, with changing landscapes enabled by connectivity and automation, future transportation networks could shift dramatically with the large-scale deployment of CAVs. With the generation of massive amounts of data from vehicles and infrastructure there are opportunities to develop optimization methods to identify and realize a substantial energy reduction of the transportation network, and to optimize the large-scale system behavior using the interplay among vehicles.

In a transportation network with CAVs there is additional information available that can be used to control and optimize jointly both vehicle-level and powertrain-level operation. For example, control technologies have been reported aimed at maximizing the energy efficiency of a 2016 Audi A3 e-tron plug-in HEV by more than 25% without degradation in tailpipe out exhaust emission levels, and without sacrificing the vehicle's drivability, performance, and safety [107]. These technologies can: (i) optimize the vehicle's speed profile aimed at minimizing (ideally, eliminating) stop-and-go driving, and (ii) optimize the powertrain of the vehicle for this optimal speed profile obtained under (i). The control architecture of such technologies (Figure 4.13) consist of (i) a vehicle dynamic (VD) controller, (ii) a powertrain (PT) controller, and (iii) a supervisory controller.

- The supervisory controller (1) oversees the VD and PT controllers, (2) communicates the endogenous and exogenous information appropriately, (3) computes the optimal routing for any desired origin-destination, (4) determines the regions where electric driving will have the major impact to derive a desired battery SOC trajectory, and (5) synthesizes a description of the upcoming road segment from the exogenous information that it communicates to the VD controller.
- The VD controller optimizes online the acceleration/deceleration and speed profile of the vehicle, and thus, the vehicle's torque demand.
- The PT controller computes the optimal nominal operation ("set-points") for the engine, motor, battery, and transmission corresponding to the optimal solution of the VD controller.

The optimal solution of the VD controller along with the endogenous and exogenous information is communicated to the PT controller through the supervisory controller. A unique feature of the control architecture is that the supervisory controller coordinates the VD and PT controllers to ensure the optimal solution yielded by the VD controller is feasible for the PT controller, and eventually results in maximization of the vehicle's energy efficiency. If the optimal solution of the VD controller is not feasible, then the supervisory controller enforces the

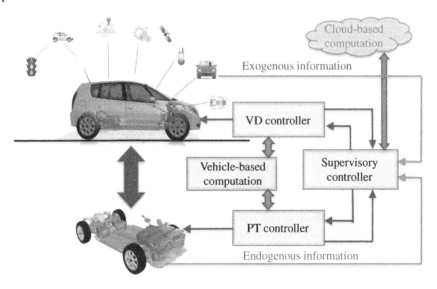

Figure 4.13 Control architecture in a connected and automated vehicle.

VD controller to repeat the optimization for a new set of parameters. To this end, we focus on approaches that have been developed for the VD controller.

4.5.1 Vehicle Dynamic Controller

In a typical commute, we encounter traffic scenarios that include crossing intersections, merging at roadways and roundabouts, cruising in congested traffic, passing through speed reduction zones, and lane-merging or passing maneuvers. There have been two major approaches to utilizing connectivity and automation to improve transportation efficiency and safety, namely, (i) platooning and (ii) traffic smoothing.

The first approach utilizes connectivity and automation to form closely-coupled vehicular platoons to reduce aerodynamic drag effectively, especially at high cruising speeds. The concept of forming platoons of vehicles traveling at a high speed was a popular system-level approach to address traffic congestion, which gained momentum in the 1980s and 1990s [108, 109]. Such automated transportation system can alleviate congestion, reduce energy use and emissions, and improve safety while increasing throughput significantly. The Japan ITS Energy Project [110], the Safe Road Trains for the Environment program [111], and the California Partner for Advanced Transportation Technology [112] are among the most reported efforts in this area.

The second approach is to smooth the traffic flow through a VD controller for optimal coordination of CAVs in different traffic scenarios, e.g, at merging roadways [113, 114], intersections [115–120], adjacent intersections [121, 122], speed reduction zones [123], roundabouts [124], and corridors [125–128]. One of the very early efforts in this direction was proposed by Athans [129] for safe and efficient coordination of merging maneuvers with the intention of avoiding congestion. Assuming a given merging sequence, the merging problem was formulated as a linear optimal regulator [130] to control a single string of vehicles, with the aim of minimizing the speed errors that will affect the desired headway between each consecutive pair of vehicles. In 2004, Dresner and Stone [131] proposed the use of the reservation scheme to control a single intersection of two roads. In their approach, each vehicle requests the reservation of the space-time cells to cross the intersection at a particular time interval defined from the estimated arrival time to the intersection. Since then, numerous approaches have been proposed on coordinating CAVs to improve traffic flow [132–134], and to achieve safe and efficient control of traffic through various traffic bottlenecks where potential vehicle collisions may happen [117, 135–142]. Queuing theory has also been used to address this problem by modeling coordination of CAVs as a polling system with two queues and one server that determines the sequence of times assigned to the vehicles on each road [143]. Some of the methods presented in the literature have focused on multi-objective optimization problems [144–148]. More recently, a study [149] indicated that transitioning from intersections with traffic lights to autonomous intersections, where vehicles can coordinate and cross the intersection without the use of traffic lights, has the potential of doubling capacity and reducing delays. Two survey papers that report the research efforts in this area can be found in [150, 151].

4.5.2 Impact and Future Directions

Several research efforts have focused on quantifying the impact of CAVs on vehicle miles traveled, energy, and greenhouse gas (GHG) emissions [152–156]. Some studies [152, 153, 155] have shown a decrease in GHG emissions with significant implications on public transit [153]. Other research efforts have investigated the feasibility and potential environmental impacts [157] of shared CAVs [158–166]. There have been also studies focusing on cost-benefit analysis of a mobility system with CAVs [105, 167–169] and the impact on vehicle ownership by using surveys or

Figure 4.14 A mixed-traffic environment consisting of human-driven and connected automated vehicles.

comparable analysis with conventional car-sharing systems [170–174]. There are several survey papers providing a good review in related topics, e.g. see [175–181].

Recent studies [182, 183], have investigated the impact of different penetration rates of CAVs, e.g. from 0% to 100%, in fuel consumption and travel time for two traffic scenarios: (i) merging at highways on ramp and (ii) merging at roundabouts. What was observed is that as the penetration rate of CAVs is decreased, both fuel consumption and travel time deteriorate. It is expected that CAVs will gradually penetrate the market, interact with human-driven vehicles (HDVs) (Figure 4.14), and contend with V2V and V2I communication limitations, e.g. bandwidth, dropouts, errors, and/or delays, as well as perception delays, lack of state information, etc. However, different levels of vehicle automation in the transportation network can significantly alter transportation efficiency metrics ranging from 45% improvement to 60% deterioration [184]. Moreover, we anticipate that efficient transportation and travel cost reduction might alter human travel behavior causing rebound effects, e.g. by improving efficiency, travel cost is decreased, hence willingness-to-travel is increased. The latter would increase overall vehicle miles traveled, which in turn might negate the benefits in terms of energy and travel time.

As we move to increasingly diverse mobility systems with different penetration rates of CAVs, new approaches are needed to optimize the impact on system behavior of the interplay between CAVs and HDVs at different traffic scenarios. While several studies have shown the benefits of CAVs to reduce energy and alleviate traffic congestion in specific traffic scenarios, most of these efforts have focused on 100% CAV penetration rates without considering HDVs. One key question that still remains unanswered is "how can CAVs and HDVs be coordinated safely?"

While several studies discussed in this chapter have shown the benefits of emerging mobility systems to reduce energy and alleviate traffic congestion in specific transportation scenarios, the research community should focus on developing a mobility system that can enhance accessibility, safety, and equity in transportation without causing rebound effects, while also gaining the travelers' acceptance. Future research should address this question and attempt to design a "socially-optimal mobility system," i.e. a mobility system that (i) is efficient (in terms of energy consumption and travel time), (ii) mitigates rebound effects, and (iii) ensures equity in transportation.

References

1 André Aronis. Control and Diagnosis of a SCR-ASC After-Treatment System for NOx and NH$_3$ Emission Reduction Under Real Driving Conditions and Potential System Failure (PhD Thesis). PhD thesis, Universitat Politècnica de València, 2023.

2 Michelle Monforte, Marie Rajon Bernard, Yoann Bernard, Georg Bieker, Kaylin Lee, Peter Mock, Jayant Mukhopadhaya, Eamonn Mulholland, Pierre-Louis Ragon, Felipe Rodriguez, Uwe Tietge, Sandra Wappelhorst, and Yuntian Zhang. European vehicle market statistics 2022/23, 2023.

3 D. Lanzarotto, M. Marchesoni, M. Passalacqua, A. Pini Prato, and M. Repetto. Overview of different hybrid vehicle architectures. *IFAC-PapersOnLine*, 51(9):218–222, 2018. 15th IFAC Symposium on Control in Transportation Systems CTS 2018.

4 Chan-Chiao Lin, Huei Peng, J.W. Grizzle, and Jun-Mo Kang. Power management strategy for a parallel hybrid electric truck. *IEEE Transactions on Control Systems Technology*, 11(6):839–849, 2003.

5 Cristian Musardo, Giorgio Rizzoni, Yann Guezennec, and Benedetto Staccia. A-ECMS: An adaptive algorithm for hybrid electric vehicle energy management. *European Journal of Control - EUR J CONTROL*, 11:509–524, 2005.

6 Lorenzo Serrao, Simona Onori, and Giorgio Rizzoni. ECMS as a realization of Pontryagin's minimum principle for HEV control. In *2009 American Control Conference*, pages 3964–3969, 2009.

7 Hoseinali Borhan, Ardalan Vahidi, Anthony Phillips, Ming Kuang, Ilya Kolmanovsky, and Stefano Di Cairano. MPC-based energy management of a power-split hybrid electric vehicle. *IEEE Transactions on Control Systems Technology*, 20:593–603, Jun 2012.

8 Hyein Jung, Tae Hoon Oh, Hyun Min Park, Heeyun Lee, and Jong Min Lee. Hybrid model predictive control for hybrid electric vehicle energy management using an efficient mixed-integer formulation. *IFAC-PapersOnLine*, 55(7):501–506, 2022.

9 Yang Zhou, Alexandre Ravey, and Marie-Cécile Péra. A survey on driving prediction techniques for predictive energy management of plug-in hybrid electric vehicles. *Journal of Power Sources*, 412:480–495, 2019.

10 Yujie Wang, Jiaqiang Tian, Zhendong Sun, Li Wang, Ruilong Xu, Mince Li, and Zonghai Chen. A comprehensive review of battery modeling and state estimation approaches for advanced battery management systems. *Renewable and Sustainable Energy Reviews*, 131:110015, 2020.

11 Johannes Hofstetter, Hans Bauer, Wenbin Li, and Georg Wachtmeister. Energy and emission management of hybrid electric vehicles using reinforcement learning. *IFAC-PapersOnLine*, 52(29):19–24, 2019. 13th IFAC Workshop on Adaptive and Learning Control Systems ALCOS 2019.

12 Ashwin Jacob and B. Ashok. An interdisciplinary review on calibration strategies of engine management system for diverse alternative fuels in IC engine applications. *Fuel*, 278:118236, 2020.

13 Dogan Erdemir and Ibrahim Dincer. A perspective on the use of ammonia as a clean fuel: Challenges and solutions. *International Journal of Energy Research*, 45(4):4827–4834, 2021.

14 Carlos Guardiola, Benjamín Pla, Pau Bares, and Alvin Barbier. Safe operation of dual-fuel engines using constrained stochastic control. *International Journal of Engine Research*, 23(2):285–299, 2022.

15 Lu Xia, Robbert Willems, Bram de Jager, and Frank Willems. Constrained optimization of fuel efficiency for RCCI engines **This work is supported by the research programme "Towards a HiEff engine" through the Netherlands Organisation for Scientific Research under STW project 14927. *IFAC-PapersOnLine*, 52(5):648–653, 2019. 9th IFAC Symposium on Advances in Automotive Control AAC 2019.

16 Hans Aulin, Thomas Johansson, Per Tunestål, and Bengt Johansson. Control of a turbo charged NVO HCCI engine using a model based approach. *IFAC Proceedings Volumes*, 42(26):79–86, 2009. 2nd IFAC Workshop on Engine and Powertrain Control, Simulation and Modeling.

17 Patrick Gorzelic, Prasad Shingne, Jason Martz, Anna Stefanopoulou, Jeff Sterniak, and Li Jiang. A low-order adaptive engine model for SI–HCCI mode transition control applications with cam switching strategies. *International Journal of Engine Research*, 17(4):451–468, 2016.

18 Kristopher Quillen, Matthew Viele, and Stephen Ciatti. Next-cycle and same-cycle cylinder pressure based control of internal combustion engines, 09 2010.

19 Carlos Jorques, Ola Stenlaas, and Per Tunestal. FPGA implementation of in-cycle closed-loop combustion control methods. 09 2021.

20 Thomas Körfer, S. Kevin Chen, Matthew Younkins, Robert Wang, Mauro Scassa, Shino George, and Marco Nencioni. Dynamic skip fire applied to a diesel engine for improved fuel consumption and emissions. In *WCX SAE World Congress Experience*. SAE International, Apr 2019.

21 Luigi del Re, Frank Allgöwer, Luigi Glielmo, Carlos Guardiola, and Ilya Kolmanovsky. *Automotive Model Predictive Control*. Springer, 2010.

22 Carlos Guardiola, Benjamín Pla, Marcelo Real, Cyril Travaillard, and Frederic Dambricourt. Fuel-to-air ratio control under short-circuit conditions through UEGO sensor signal analysis. *International Journal of Engine Research*, 21(9):1577–1583, 2020.

23 Hsiu-Ming Wu and Reza Tafreshi. Observer-based internal model air–fuel ratio control of lean-burn si engines. *IFAC Journal of Systems and Control*, 9:100065, 2019.

24 Stijn van Dooren, Alois Amstutz, and Christopher H. Onder. A causal supervisory control strategy for optimal control of a heavy-duty diesel engine with SCR aftertreatment. *Control Engineering Practice*, 119:104982, 2022.

25 M. Aliramezani, C. R. Koch, and R. E. Hayes. Estimating tailpipe NOx concentration using a dynamic NOx/ammonia cross sensitivity model coupled to a three state control oriented SCR model. *IFAC-PapersOnLine*, 49(11):8–13, 2016. 8th IFAC Symposium on Advances in Automotive Control AAC 2016.

26 Benjamín Pla, Pedro Piqueras, Pau Bares, and André Aronis. NO_x sensor cross sensitivity model and simultaneous prediction of NO_x and NH_3 slip from automotive catalytic converters under real driving conditions. *International Journal of Engine Research*, 22(10):3209–3218, 2021.

27 Carlos Guardiola, Benjamin Pla, Pau Bares, and Javier Mora. An on-board method to estimate the light-off temperature of diesel oxidation catalysts. *International Journal of Engine Research*, 21(8):1480–1492, 2020.

28 Carlos Guardiola, Benjamin Pla, Javier Mora Pérez, and Damien Lefebvre. Experimental determination and modelling of the diesel oxidation catalysts ageing effects. *Proceedings of the Institution of Mechanical Engineers, Part D: Journal of Automobile Engineering*, 233(12):3016–3029, 2019.

29 Rasoul Salehi and Anna Stefanopoulou. Optimal exhaust valve opening control for fast aftertreatment warm up in diesel engines. In *ASME 2018 Dynamic Systems and Control Conference*, Oct. 2018.

30 Futoshi Nakano and Christopher Eck. Robust DPF regeneration control for cost-effective small commercial vehicles. In *13th International Conference on Engines & Vehicles*. SAE International, Sep. 2017.

31 Anyu Cheng, Haining Li, and Liangbo Xiong. A model-based calibration method of automotive electronic control unit. *Wuhan University Journal of Natural Sciences*, 21:171–177, 2016. doi: 10.1007/s11859-016-1155-8.

32 Ivan Arsie, Andrea Cricchio, Matteo De Cesare, Francesco Lazzarini, Cesare Pianese, and Marco Sorrentino. Neural network models for virtual sensing of NOx emissions in automotive diesel engines with least square-based adaptation. *Control Engineering Practice*, 61:11–20, 2017.

33 Prasoon Garg, Emilia Silvas, and Frank Willems. Potential of machine learning methods for robust performance and efficient engine control development. *IFAC-PapersOnLine*, 54(10):189–195, 2021. 6th IFAC Conference on Engine Powertrain Control, Simulation and Modeling E-COSM 2021.

34 Carlos Guardiola, Christian Vigild, Frederik De Smet, and Klaus Schusteritz. From OBD to connected diagnostics: A game changer at fleet, vehicle and component level**Ford-UPV collaboration on connected diagnostics is funded by EU H2020 ICT-56-2020 grant 957258 ASSIST-IoT. C Guardiola also acknowledges Spanish Agencia Estatal de Investigación grant PID2019-108031RB-C21. *IFAC-PapersOnLine*, 54(10):558–563, 2021. 6th IFAC Conference on Engine Powertrain Control, Simulation and Modeling E-COSM 2021.

35 Vaishnavi Karanam, Adam Davis, Claire Sugihara, Katrina Sutton, and Gil Tal. From shifting gears to changing modes: The impact of driver inputs on plug-in hybrid electric vehicle energy use & emissions. *Transportation Research Interdisciplinary Perspectives*, 14:100597, 2022.

36 Yiqun Liu, Y. Gene Liao, and Ming-Chia Lai. Fuel economy improvement and emission reduction of 48 V mild hybrid electric vehicles with P0, P1, and P2 architectures with lithium battery cell experimental data. *Advances in Mechanical Engineering*, 13(10):168781402110360. doi: 10.1177/16878140211036022.

37 Steve Hayslett, Keith Van Maanen, Wolfgang Wenzel, and Tausif Husain. The 48-V mild hybrid: Benefits, motivation, and the future outlook. *IEEE Electrification Magazine*, 8(2):11–17. doi: 10.1109/mele.2020.2985481.

38 Shima Nazari, Jason Siegel, Robert Middleton, and Anna Stefanopoulou. Power split supercharging: A mild hybrid approach to boost fuel economy. *Energies*, 13(24):6580. doi: 10.3390/en13246580.

39 Iqbal Husain, Burak Ozpineci, Md Sariful Islam, Emre Gurpinar, Gui-Jia Su, Wensong Yu, Shajjad Chowdhury, Lincoln Xue, Dhrubo Rahman, and Raj Sahu. Electric drive technology trends, challenges, and opportunities for future electric vehicles. *Proceedings of the IEEE*, 109(6):1039–1059. doi: 10.1109/jproc.2020.3046112.

40 Yuanfeng Lan, Yassine Benomar, Kritika Deepak, Ahmet Aksoz, Mohamed El Baghdadi, Emine Bostanci, and Omar Hegazy. Switched reluctance motors and drive systems for electric vehicle powertrains: State of the art analysis and future trends. *Energies*, 14(8):2079. doi: 10.3390/en14082079.

41 P. Ramesh, M. Umavathi, C. Bharatiraja, G. Ramanathan, and Sivaprasad Athikkal. Development of a PMSM motor field-oriented control algorithm for electrical vehicles. *Materials Today: Proceedings*, 65:176–187. doi: 10.1016/j.matpr.2022.06.080.

42 Deepak Mohanraj, Janaki Gopalakrishnan, Bharatiraja Chokkalingam, and Lucian Mihet-Popa. Critical aspects of electric motor drive controllers and mitigation of torque ripple—review. *IEEE Access*, 10:73635–73674. doi: 10.1109/ACCESS.2022.3187515.

43 Jiyu Zhang, Hongyang Yao, and Giorgio Rizzoni. Fault diagnosis for electric drive systems of electrified vehicles based on structural analysis. *IEEE Transactions on Vehicular Technology*, 66(2):1027–1039, 2017. doi: 10.1109/TVT.2016.2556691.

44 Anna Tomaszewska, Zhengyu Chu, Xuning Feng, Simon O'Kane, Xinhua Liu, Jingyi Chen, Chenzhen Ji, Elizabeth Endler, Ruihe Li, Lishuo Liu, Yalun Li, Siqi Zheng, Sebastian Vetterlein, Ming Gao, Jiuyu Du, Michael Parkes, Minggao Ouyang, Monica Marinescu, Gregory Offer, and Billy Wu. Lithium-ion battery fast charging: A review. *eTransportation*, 1:100011. doi: 10.1016/j.etran.2019.100011.

45 Xuning Feng, Minggao Ouyang, Xiang Liu, Languang Lu, Yong Xia, and Xiangming He. Thermal runaway mechanism of lithium ion battery for electric vehicles: A review. *Energy Storage Materials*, 10:246–267. doi: 10.1016/j.ensm.2017.05.013.

46 Qibo Li, Chuanbo Yang, Shriram Santhanagopalan, Kandler Smith, Joshua Lamb, Leigh Anna Steele, and Loraine Torres-Castro. Numerical investigation of thermal runaway mitigation through a passive thermal management system. *Journal of Power Sources*, 429:80–88. doi: 10.1016/j.jpowsour.2019.04.091.

47 Ting Cai, Anna G. Stefanopoulou, and Jason B. Siegel. Modeling li-ion battery temperature and expansion force during the early stages of thermal runaway triggered by internal shorts. *Journal of the Electrochemical Society*, 166(12):A2431–A2443. doi: 10.1149/2.1561910jes.

48 G.L. Plett. *Battery Management Systems*, volume 1, Artech House power engineering and power electronics. Artech House. ISBN 9781630810238.

49 Johannes Schmalstieg, Christiane Rahe, Madeleine Ecker, and Dirk Uwe Sauer. Full cell parameterization of a high-power lithium-ion battery for a physico-chemical model: Part I. Physical and electrochemical parameters. *Journal of the Electrochemical Society*, 165(16):A3799–A3810, 2018. doi: 10.1149/2.0321816jes.

50 Jackson Fogelquist, Qingzhi Lai, and Xinfan Lin. On the error of li-ion battery parameter estimation subject to system uncertainties. *Journal of the Electrochemical Society*, 170(3):030510, 2023. doi: 10.1149/1945-7111/acbc9c.

51 Xinfan Lin, Hector E. Perez, Jason B. Siegel, Anna G. Stefanopoulou, Yonghua Li, R. Dyche Anderson, Yi Ding, and Matthew P. Castanier. Online parameterization of lumped thermal dynamics in cylindrical lithium ion batteries for core temperature estimation and health monitoring. *IEEE Transactions on Control Systems Technology*, 21(5):1745–1755, 2013. doi: 10.1109/TCST.2012.2217143.

52 Adrien M. Bizeray, Jin-Ho Kim, Stephen R. Duncan, and David A. Howey. Identifiability and parameter estimation of the single particle lithium-ion battery mode. *IEEE Transactions on Control Systems Technology*, 27(5):1862–1877, 2019. doi: 10.1109/TCST.2018.2838097.

53 Thomas F. Fuller, Marc Doyle, and John Newman. Relaxation phenomena in lithium-ion-insertion cells. *Journal of the Electrochemical Society*, 141(4):982–990. doi: 10.1149/1.2054868.

54 Zhiqiang Chen, Dmitri L. Danilov, Rüdiger-A. Eichel, and Peter H. L. Notten. Porous electrode modeling and its applications to li-ion batteries. *Advanced Energy Materials*, 12(32):2201506. doi: 10.1002/aenm.202201506.

55 Scott G. Marquis, Valentin Sulzer, Robert Timms, Colin P. Please, and S. Jon Chapman. An asymptotic derivation of a single particle model with electrolyte. *Journal of the Electrochemical Society*, 166(15):A3693–A3706. doi: 10.1149/2.0341915jes.

56 Krishnakumar Gopalakrishnan and Gregory J. Offer. A composite single particle lithium-ion battery model through system identification. *IEEE Transactions on Control Systems Technology*, 30(1):1–13. doi: 10.1109/TCST.2020.3047776.

57 Luis D. Couto, Raffaele Romagnoli, Saehong Park, Dong Zhang, Scott J. Moura, Michel Kinnaert, and Emanuele Garone. Faster and healthier charging of lithium-ion batteries via constrained feedback control. *IEEE Transactions on Control Systems Technology*, 30(5):1990–2001. doi: 10.1109/tcst.2021.3135149.

58 Valentin Sulzer, Peyman Mohtat, Antti Aitio, Suhak Lee, Yen T. Yeh, Frank Steinbacher, Muhammad Umer Khan, Jang Woo Lee, Jason B. Siegel, Anna G. Stefanopoulou, and David A. Howey. The challenge and opportunity of battery lifetime prediction from field data. *Joule*, 5(8):1934–1955. doi: 10.1016/j.joule.2021.06.005.

59 Suhak Lee, Jason B. Siegel, Anna G. Stefanopoulou, Jang-Woo Lee, and Tae-Kyung Lee. Electrode state of health estimation for lithium ion batteries considering half-cell potential change due to aging. *Journal of the Electrochemical Society*, 167(9):090531. doi: 10.1149/1945-7111/ab8c83.

60 Caitlin Murphy, Trieu Mai, Yinong Sun, Paige Jadun, Matteo Muratori, Brent Nelson, and Ryan Jones. *Electrification Futures Study: Scenarios of Power System Evolution and Infrastructure Development for the United States*. National Renewable Energy Laboratory, Golden, CO. NREL/TP-6A20-72330. https://www.nrel.gov/docs/fy21osti/72330.pdf, 2021.

61 Liren Yang, Amey Karnik, Benjamin Pence, Md Tawhid Bin Waez, and Necmiye Ozay. Fuel cell thermal management: Modeling, specifications, and correct-by-construction control synthesis. *IEEE Transactions on Control Systems Technology*, 28(5):1638–1651. doi: 10.1109/TCST.2019.2918747.

62 S. Vengatesan, Michael W. Fowler, Xiao-Zi Yuan, and Haijiang Wang. Diagnosis of MEA degradation under accelerated relative humidity cycling. *Journal of Power Sources*, 196(11):5045–5052. doi: 10.1016/j.jpowsour.2011.01.088.

63 Jinfeng Wu, Xiao Zi Yuan, Jonathan J. Martin, Haijiang Wang, Jiujun Zhang, Jun Shen, Shaohong Wu, and Walter Merida. A review of PEM fuel cell durability: Degradation mechanisms and mitigation strategies. *Journal of Power Sources*, 184(1):104–119. doi: 10.1016/j.jpowsour.2008.06.006.

64 Jay T. Pukrushpan, Anna G. Stefanopoulou, and Huei Peng. *Control of Fuel Cell Power Systems*. doi: 10.1007/978-1-4471-3792-4.

65 Tong Zhang, Peiqi Wang, Huicui Chen, and Pucheng Pei. A review of automotive proton exchange membrane fuel cell degradation under start-stop operating condition. *Applied Energy*, 223:249–262. doi: 10.1016/j.apenergy.2018.04.049.

66 Brian D. James. Fuel cell cost and performance analysis. *2022 DOE Hydrogen and Fuel Cells Program Annual Merit Review and Peer Evaluation Meeting Presentation.*

67 SAE International. Taxonomy and Definitions for Terms Related to Driving Automation Systems for On-Road Motor Vehicles. SAE International, 2021.

68 Hongtei Eric Tseng, Behrouz Ashrafi, Dinu Madau, T. Allen Brown, and Darrel Recker. The development of vehicle stability control at ford. *IEEE/ASME Transactions on Mechatronics*, 4(3):223–234, 1999.

69 Lingyun Xiao and Feng Gao. A comprehensive review of the development of adaptive cruise control systems. *Vehicle system dynamics*, 48(10):1167–1192, 2010.

70 Angela H. Eichelberger and Anne T. McCartt. Toyota drivers' experiences with dynamic radar cruise control, pre-collision system, and lane-keeping assist. *Journal of Safety Research*, 56:67–73, 2016.

71 Andrea Censi, Konstantin Slutsky, Tichakorn Wongpiromsarn, Dmitry Yershov, Scott Pendleton, James Fu, and Emilio Frazzoli. Liability, ethics, and culture-aware behavior specification using rulebooks. In *International Conference on Robotics and Automation*, pages 8536–8542, 2019.

72 Shreejith Ravikumar, Rien Quirynen, Akshay Bhagat, Eyad Zeino, and Stefano Di Cairano. Mixed-integer programming for centralized coordination of connected and automated vehicles in dynamic environment. In *IEEE Conference on Control Technology and Applications*, pages 814–819, 2021.

73 Zejiang Wang, Ahmad Ahmad, Rien Quirynen, Yebin Wang, Akshay Bhagat, Eyad Zeino, Yuji Zushi, and Stefano Di Cairano. Motion planning and model predictive control for automated tractor-trailer hitching maneuver. In *IEEE Conference on Control Technology and Applications*, pages 676–682, 2022.

74 Karl J. Åström. Automatic control—the hidden technology. In *Advances in Control*, pages 1–28. Springer, 1999.

75 Martin Buehler, Karl Iagnemma, and Sanjiv Singh. *The 2005 DARPA Grand Challenge: The Great Robot Race*, volume 36. Springer, 2007.

76 M. Buehler, K. Iagnemma, and S. Singh. *The DARPA Urban Challenge: Autonomous Vehicles in City Traffic*, volume 56. Springer, 2009.

77 Karl Berntorp, Tru Hoang, Rien Quirynen, and Stefano Di Cairano. Control architecture design for autonomous vehicles. In *IEEE Conference on Control Technology and Applications*, pages 404–411, 2018.

78 Heejin Ahn, Karl Berntorp, Pranav Inani, Arjun Jagdish Ram, and Stefano Di Cairano. Reachability-based decision-making for autonomous driving: Theory and experiments. *IEEE Transactions on Control Systems Technology*, 29(5):1907–1921, 2020.

79 Heejin Ahn, Karl Berntorp, and Stefano Di Cairano. Cooperating modular goal selection and motion planning for autonomous driving. In *59th IEEE Conference on Decision and Control*, pages 3481–3486, 2020.

80 Karl Berntorp, Richard Bai, Karl F. Erliksson, Claus Danielson, Avishai Weiss, and Stefano Di Cairano. Positive invariant sets for safe integrated vehicle motion planning and control. *IEEE Trans. Intelligent Vehicles*, 5(1):112–126, 2019.

81 Mitsubishi Electric Co. Mitsubishi electric key growth businesses: Automotive equipment (xEV/ADAS). https://www.mitsubishielectric.com/news/2021/pdf/1109-b.pdf, 2021. Accessed: 2022-12-13.

82 François Chaumette, Seth Hutchinson, and Peter Corke. Visual servoing. *Springer Handbook of Robotics*, pages 841–866. Springer, 2016.

83 Christoforos Kanellakis and George Nikolakopoulos. Survey on computer vision for UAVs: Current developments and trends. *Journal of Intelligent & Robotic Systems*, 87:141–168, 2017.

84 Takafumi Taketomi, Hideaki Uchiyama, and Sei Ikeda. Visual SLAM algorithms: A survey from 2010 to 2016. *IPSJ Transactions on Computer Vision and Applications*, 9(1):1–11, 2017.

85 Davide Falanga, Philipp Foehn, Peng Lu, and Davide Scaramuzza. PAMPC: Perception-aware model predictive control for quadrotors. In *2018 IEEE/RSJ International Conference on Intelligent Robots and Systems*, pages 1–8, 2018.

86 Kemin Zhou and John Comstock Doyle. *Essentials of Robust Control*, volume 104. Prentice Hall, Upper Saddle River, NJ, 1998.

87 James Blake Rawlings and David Q. Mayne. *Model Predictive Control: Theory and Design*. Nob Hill Pub., 2009.

88 Davor Hrovat, Stefano Di Cairano, H. Eric Tseng, and Ilya V. Kolmanovsky. The development of model predictive control in automotive industry: A survey. In *Proceedings of IEEE International Conference on Control Applications*, pages 295–302, 2012.

89 Stefano Di Cairano and Ilya V. Kolmanovsky. Real-time optimization and model predictive control for aerospace and automotive applications. In *American Control Conference*, pages 2392–2409, 2018.

90 Stefano Di Cairano, Hongtei Eric Tseng, Daniele Bernardini, and Alberto Bemporad. Vehicle yaw stability control by coordinated active front steering and differential braking in the tire sideslip angles domain. *IEEE Transactions on Control Systems Technology*, 21(4):1236–1248, 2012.

91 Aaron D. Ames, Jessy W. Grizzle, and Paulo Tabuada. Control barrier function based quadratic programs with application to adaptive cruise control. In *Proceedings of the 53rd IEEE Conference Decision and Control*, pages 6271–6278, 2014.

92 Stefano Di Cairano, U.V. Kalabić, and Karl Berntorp. Vehicle tracking control on piecewise-clothoidal trajectories by MPC with guaranteed error bounds. In *Proceedings of the 55th IEEE Conference on Decision and Control*, pages 709–714, 2016.

93 Stefano Di Cairano, Tobias Bäthge, and Rolf Findeisen. Modular design for constrained control of actuator-plant cascades. In *Proceedings of American Control Conference*, pages 1755–1760, 2019.

94 Sebastian Brechtel, Tobias Gindele, and Rüdiger Dillmann. Probabilistic decision-making under uncertainty for autonomous driving using continuous POMDPs. In *17th International IEEE Conference on Intelligent Transportation Systems*, pages 392–399, 2014.

95 Changxi You, Jianbo Lu, Dimitar Filev, and Panagiotis Tsiotras. Highway traffic modeling and decision making for autonomous vehicle using reinforcement learning. In *IEEE Intelligent Vehicles Symposium*, pages 1227–1232, 2018.

96 Carl-Johan Hoel, Katherine Driggs-Campbell, Krister Wolff, Leo Laine, and Mykel J. Kochenderfer. Combining planning and deep reinforcement learning in tactical decision making for autonomous driving. *IEEE Transactions on Intelligent Vehicles*, 5(2):294–305, 2019.

97 Engin Ozatay, Simona Onori, James Wollaeger, Umit Ozguner, Giorgio Rizzoni, Dimitar Filev, John Michelini, and Stefano Di Cairano. Cloud-based velocity profile optimization for everyday driving: A dynamic-programming-based solution. *IEEE Trans Intelligent Transportation Systems*, 15(6):2491–2505, 2014.

98 Gilsoo Lee, Jianlin Guo, Kyeong Jin Kim, Philip Orlik, Heejin Ahn, Stefano Di Cairano, and Walid Saad. Edge computing for interconnected intersections in internet of vehicles. In *IEEE Intelligent Vehicles Symposium*, pages 480–486, 2020.

99 Tengchan Zeng, Jianlin Guo, Kyeong Jin Kim, Kieran Parsons, Philip Orlik, Stefano Di Cairano, and Walid Saad. Multi-task federated learning for traffic prediction and its application to route planning. In *IEEE Intelligent Vehicles Symposium*, pages 451–457, 2021.

100 Marcel Menner, Ziyi Ma, Karl Berntorp, and Stefano Di Cairano. Location and driver-specific vehicle adaptation using crowdsourced data. In *Proceedings of European Control Conference*, pages 769–774, 2022.

101 Angelo Domenico Bonzanini, Ali Mesbah, and Stefano Di Cairano. Perception-aware chance-constrained model predictive control for uncertain environments. In *Proceedings of the American Control Conference*, 2021.

102 Angelo Domenico Bonzanini, Ali Mesbah, and Stefano Di Cairano. Multi-stage perception-aware chance-constrained MPC with application to automated driving. In *Proceedings of the American Control Conference*, 2022.

103 Les Hatton. Safer language subsets: An overview and a case history, MISRA C. *Information and Software Technology*, 46(7):465–472, 2004.

104 Kevin Spieser, Kyle Treleaven, Rick Zhang, Emilio Frazzoli, Daniel Morton, and Marco Pavone. Toward a systematic approach to the design and evaluation of automated mobility-on-demand systems: A case study in Singapore. In *Road vehicle automation*, pages 229–245. Springer, 2014.

105 Daniel J. Fagnant and Kara M. Kockelman. The travel and environmental implications of shared autonomous vehicles, using agent-based model scenarios. *Transportation Research Part C: Emerging Technologies*, 40:1–13, 2014.

106 Andreas A. Malikopoulos. A duality framework for stochastic optimal control of complex systems. *IEEE Transactions on Automatic Control*, 61(10):2756–2765, 2016.

107 A.M. Ishtiaque Mahbub and Andreas A. Malikopoulos. Concurrent optimization of vehicle dynamics and powertrain operation using connectivity and automation. In SAE Technical Paper 2020-01-0580. SAE International, 2020. doi: 10.4271/2020-01-0580.

108 S.E. Shladover, C.A. Desoer, J.K. Hedrick, M. Tomizuka, J. Walrand, W.-B. Zhang, D.H. McMahon, H. Peng, S. Sheikholeslam, and N. McKeown. Automated vehicle control developments in the PATH program. *IEEE Transactions on Vehicular Technology*, 40(1):114–130, 1991.

109 R. Rajamani, H.-S. Tan, B.K. Law, and W.-B. Zhang. Demonstration of integrated longitudinal and lateral control for the operation of automated vehicles in platoons. *IEEE Transactions on Control Systems Technology*, 8(4):695–708, 2000.

110 Sadayuki Tsugawa. An overview on an automated truck platoon within the energy its project. *IFAC Proceedings Volumes*, 46(21):41–46, 2013.

111 Arturo Dávila and Mario Nombela. SARTRE: Safe road trains for the environment. In *Conference on Personal Rapid Transit PRT@LHR*, volume 3, pages 2–3, 2010.

112 Steven E Shladover. PATH at 20–history and major milestones. *IEEE Transactions on Intelligent Transportation Systems*, 8(4):584–592, 2007.

113 Jackeline Rios-Torres and Andreas A. Malikopoulos. Automated and cooperative vehicle merging at highway on-ramps. *IEEE Transactions on Intelligent Transportation Systems*, 18(4):780–789, 2017.

114 Ioannis A. Ntousakis, Ioannis K. Nikolos, and Markos Papageorgiou. Optimal vehicle trajectory planning in the context of cooperative merging on highways. *Transportation Research Part C: Emerging Technologies*, 71:464–488, 2016.

115 Andreas A. Malikopoulos, Christos G. Cassandras, and Yue J. Zhang. A decentralized energy-optimal control framework for connected automated vehicles at signal-free intersections. *Automatica*, 93:244–256, 2018.

116 Andreas A. Malikopoulos, Logan E. Beaver, and Ioannis Vasileios Chremos. Optimal time trajectory and coordination for connected and automated vehicles. *Automatica*, 125:109469, 2021.

117 Kurt Dresner and Peter Stone. A multiagent approach to autonomous intersection management. *Journal of Artificial Intelligence Research*, 31:591–653, 2008.

118 Joyoung Lee and Byungkyu Park. Development and evaluation of a cooperative vehicle intersection control algorithm under the connected vehicles environment. *IEEE Transactions on Intelligent Transportation Systems*, 13(1):81–90, 2012.

119 Jean Gregoire, Silvere Bonnabel, and Arnaud De La Fortelle. Priority-based intersection management with kinodynamic constraints. In *2014 European Control Conference (ECC)*, pages 2902–2907. IEEE, 2014.

120 Seyed Alireza Fayazi and Ardalan Vahidi. Mixed-integer linear programming for optimal scheduling of autonomous vehicle intersection crossing. *IEEE Transactions on Intelligent Vehicles*, 3(3):287–299, 2018.

121 Behdad Chalaki and Andreas A. Malikopoulos. Optimal control of connected and automated vehicles at multiple adjacent intersections. *IEEE Transactions on Control Systems Technology*, 30(3):972–984, 2022. doi: 10.1109/TCST.2021.3082306.

122 Behdad Chalaki and Andreas A. Malikopoulos. Time-optimal coordination for connected and automated vehicles at adjacent intersections. *IEEE Transactions on Intelligent Transportation Systems*, 23(8):13330–13345, 2021. doi: 10.1109/TITS.2021.3123479.

123 A.A. Malikopoulos, S. Hong, B. Park, J. Lee, and Seunghan Ryu. Optimal control for speed harmonization of automated vehicles. *IEEE Transactions on Intelligent Transportation Systems*, 20(7):2405–2417, 2018.

124 Behdad Chalaki, Logan E. Beaver, and Andreas A. Malikopoulos. Experimental validation of a real-time optimal controller for coordination of CAVs in a multi-lane roundabout. In *31st IEEE Intelligent Vehicles Symposium (IV)*, pages 504–509, 2020.

125 L. Zhao and Andreas A. Malikopoulos. Decentralized optimal control of connected and automated vehicles in a corridor. In *2018 21st International Conference on Intelligent Transportation Systems (ITSC)*, pages 1252–1257, Nov. 2018. doi: 10.1109/ITSC.2018.8569229.

126 A.M. Ishtiaque Mahbub, Andreas A. Malikopoulos, and Liuhui Zhao. Decentralized optimal coordination of connected and automated vehicles for multiple traffic scenarios. *Automatica*, 117:108958, 2020.

127 Liuhui Zhao, Andreas A. Malikopoulos, and Jackeline Rios-Torres. On the traffic impacts of optimally controlled connected and automated vehicles. In *Proceedings of 2019 IEEE Conference on Control Technology and Applications (CCTA)*, pages 882–887. IEEE, 2019.

128 Behdad Chalaki, Logan E. Beaver, A. M. Ishtiaque Mahbub, Heeseung Bang, and Andreas A. Malikopoulos. A research and educational robotic testbed for real-time control of emerging mobility systems: From theory to scaled experiments. *IEEE Control Systems Magazine*, 42(6):20–34, 2022.

129 Michael Athans. A unified approach to the vehicle-merging problem. *Transportation Research*, 3(1):123–133, 1969. ISSN 00411647. doi: 10.1016/0041-1647(69)90109-9.

130 W.S. Levine and M. Athans. On the optimal error regulation of a string of moving vehicles. *IEEE Transactions on Automatic Control*, 11(3):355–361, 1966.

131 K. Dresner and P. Stone. Multiagent traffic management: A reservation-based intersection control mechanism. In *Proceedings of the Third International Joint Conference on Autonomous Agents and Multiagents Systems*, pages 530–537, 2004.

132 P. Kachroo and Z. Li. Vehicle merging control design for an automated highway system. In *Proceedings of Conference on Intelligent Transportation Systems*, pages 224–229, 1997. ISBN 0-7803-4269-0.

133 M. Antoniotti, A. Deshpande, and A. Girault. Microsimulation analysis of automated vehicles on multiple merge junction highways. In *IEEE International Conference in Systems, Man, and Cybernetics*, pages 839–844, 1997.

134 B. Ran, S. Leight, and B. Chang. A microscopic simulation model for merging control on a dedicated-lane automated highway system. *Transportation Research Part C: Emerging Technologies*, 7(6):369–388, 1999.

135 Arnaud de La Fortelle. Analysis of reservation algorithms for cooperative planning at intersections. *13th International IEEE Conference on Intelligent Transportation Systems*, pages 445–449, September 2010.

136 Shan Huang, A.W. Sadek, and Yunjie Zhao. Assessing the mobility and environmental benefits of reservation-based intelligent intersections using an integrated simulator. *IEEE Transactions on Intelligent Transportation Systems*, 13(3):1201–1214, 2012.

137 Ismail H. Zohdy, Raj Kishore Kamalanathsharma, and Hesham Rakha. Intersection management for autonomous vehicles using iCACC. In *2012 15th International IEEE Conference on Intelligent Transportation Systems*, pages 1109–1114, 2012.

138 Fei Yan, Mahjoub Dridi, and Abdellah El Moudni. Autonomous vehicle sequencing algorithm at isolated intersections. In *2009 12th International IEEE Conference on Intelligent Transportation Systems*, pages 1–6, 2009.

139 Li Li and Fei-Yue Wang. Cooperative driving at blind crossings using intervehicle communication. *IEEE Transactions in Vehicular Technology*, 55(6):1712–1724, 2006. doi: 10.1109/TVT.2006.878730.

140 Feng Zhu and Satish V. Ukkusuri. A linear programming formulation for autonomous intersection control within a dynamic traffic assignment and connected vehicle environment. *Transportation Research Part C: Emerging Technologies*, 55:363–378, 2015.

141 J. Wu, F. Perronnet, and A. Abbas-Turki. Cooperative vehicle-actuator system: A sequence-based framework of cooperative intersections management. *IET Intelligent Transport Systems*, 8(4):352–360, 2014.

142 Kyoung-Dae Kim and P.R. Kumar. An MPC-based approach to provable system-wide safety and liveness of autonomous ground traffic. *IEEE Transactions on Automatic Control*, 59(12):3341–3356, 2014.

143 David Miculescu and Sertac Karaman. Polling-systems-based autonomous vehicle coordinattion in traffic intersections with no traffic signals. *IEEE Transactions on Automatic Control*, 65(2):680–694, 2020.

144 M.A.S. Kamal, M. Mukai, J. Murata, and T. Kawabe. Model predictive control of vehicles on urban roads for improved fuel economy. *IEEE Transactions on Control Systems Technology*, 21(3):831–841, 2013.

145 M.A.S. Kamal, J. Imura, T. Hayakawa, A. Ohata, and K. Aihara. A vehicle-intersection coordination scheme for smooth flows of traffic without using traffic lights. *IEEE Transactions on Intelligent Transportation Systems*, 16(3):1136–1147, 2014. doi: 10.1109/TITS.2014.2354380.

146 G. R. Campos, P. Falcone, H. Wymeersch, R. Hult, and J. Sjoberg. Cooperative receding horizon conflict resolution at traffic intersections. In *2014 IEEE 53rd Annual Conference on Decision and Control (CDC)*, pages 2932–2937, 2014.

147 Laleh Makarem, Denis Gillet, and Senior Member. Model predictive coordination of autonomous vehicles crossing intersections. In *16th International IEEE Conference on Intelligent Transportation Systems*, pages 1799–1804, 2013.

148 Xiangjun Qian, Jean Gregoire, Arnaud De La Fortelle, and Fabien Moutarde. Decentralized model predictive control for smooth coordination of automated vehicles at intersection. In *2015 European Control Conference (ECC)*, pages 3452–3458. IEEE, 2015.

149 Remi Tachet, Paolo Santi, Stanislav Sobolevsky, Luis Ignacio Reyes-Castro, Emilio Frazzoli, Dirk Helbing, and Carlo Ratti. Revisiting street intersections using slot-based systems. *PLoS ONE*, 11(3):e0149607, 2016.

150 Jackeline Rios-Torres and Andreas A. Malikopoulos. A survey on coordination of connected and automated vehicles at intersections and merging at highway on-ramps. *IEEE Transactions on Intelligent Transportation Systems*, 18(5):1066–1077, 2017.

151 Jacopo Guanetti, Yeojun Kim, and Francesco Borrelli. Control of connected and automated vehicles: State of the art and future challenges. *Annual Reviews in Control*, 45:18–40, 2018.

152 Elliot Martin and Susan A. Shaheen. Greenhouse gas emissions impacts of carsharing in North America. *Transactions on Intelligent Transportation Systems*, 12(4):1–114, 2011. ISSN 1524-9050.

153 Elliot Martin and Susan Shaheen. The impact of carsharing on public transit and non-motorized travel: An exploration of North American carsharing survey data. *Energies*, 4(11):2094–2114, 2011. ISSN 19961073.

154 Jörg Firnkorn and Martin Müller. Free-floating electric carsharing-fleets in smart cities: The dawning of a post-private car era in urban environments? *Environmental Science & Policy*, 45:30–40, 2015.

155 Elliot Martin and Susan Shaheen. Impacts of Car2Go on vehicle ownership, modal shift, vehicle miles traveled, and greenhouse gas emissions: An analysis of five North American cities. Transportation Sustainability Research Center, UC Berkeley, 2016.

156 T. Donna Chen and Kara M. Kockelman. Carsharing's life-cycle impacts on energy use and greenhouse gas emissions. *Transportation Research Part D: Transport and Environment*, 47:276–284, 2016.

157 Susan Shaheen and Michael Galczynski. Autonomous carsharing/taxi pathways. In *TRB Automated Vehicles Symposium*, 2014.

158 Z.J. Chong, Baoxing Qin, Tirthankar Bandyopadhyay, Tichakorn Wongpiromsarn, Brice Rebsamen, P. Dai, E.S. Rankin, and Marcelo H. Ang. Autonomy for mobility on demand. In *Intelligent Autonomous Systems 12*, pages 671–682. Springer, 2013.

159 Hillary Jeanette Ford. *Shared Autonomous Taxis: Implementing An Efficient Alternative to Automobile Dependency*. Princeton University, 2012.

160 Pierre-Jean Rigole. Study of a shared autonomous vehicles based mobility solution in Stockholm, 2014.

161 Joschka Bischoff and Michal Maciejewski. Simulation of city-wide replacement of private cars with autonomous taxis in Berlin. *Procedia Computer Science*, 83:237–244, 2016.

162 Hussein Dia and Farid Javanshour. Autonomous shared mobility-on-demand: Melbourne pilot simulation study. *Transportation Research Procedia*, 22:285–296, 2017.

163 Louis A. Merlin. Comparing automated shared taxis and conventional bus transit for a small city. *Journal of Public Transportation*, 20(2):19–39, 2017.

164 David Metz. Developing policy for urban autonomous vehicles: Impact on congestion. *Urban Science*, 2(2):33, 2018.

165 David Fiedler, Michal Cap, and Michal Certicky. Impact of mobility-on-demand on traffic congestion: Simulation-based study. In *2017 IEEE 20th International Conference on Intelligent Transportation Systems (ITSC)*, pages 1–6. IEEE, 2018.

166 Miaojia Lu, Morteza Taiebat, Ming Xu, and Shu-Chien Hsu. Multiagent spatial simulation of autonomous taxis for urban commute: Travel economics and environmental impacts. *Journal of Urban Planning and Development*, 144(4), 2018. doi: 10.1061/(ASCE)UP.1943-5444.0000469.

167 Lawrence D. Burns, William C. Jordan, Jordan Analytics, and Bonnie A. Scarborough. Transforming Personal Mobility. Technical report, The Earth Institute, Columbia University, 2012.

168 Patrick M. Bösch, Felix Becker, Henrik Becker, and Kay W. Axhausen. Cost-based analysis of autonomous mobility services. *Transport Policy*, 64:76–91, 2017.

169 Aditi Moorthy, Robert De Kleine, Gregory Keoleian, Jeremy Good, and Geoff Lewis. Shared autonomous vehicles as a sustainable solution to the last mile problem: A case study of Ann Arbor-Detroit area. *SAE International Journal of Passenger Cars - Electronic and Electrical Systems*, 10(2):2017–01–1276, 2017.

170 Long T. Truong, Chris De Gruyter, Graham Currie, and Alexa Delbosc. Estimating the trip generation impacts of autonomous vehicles on car travel in Victoria, Australia. *Transportation*, 44(6):1279–1292, 2017.

171 Iis P. Tussyadiah, Florian J. Zach, and Jianxi Wang. Attitudes toward autonomous on demand mobility system: The case of self-driving taxi. In *Information and Communication Technologies in Tourism 2017*, pages 755–766. Springer, 2017.

172 Dávid Földes and Csaba Csiszár. Framework for planning the mobility service based on autonomous vehicles. In *2018 Smart City Symposium Prague (SCSP)*, pages 1–6. IEEE, 2018.

173 Mingyang Hao and Toshiyuki Yamamoto. Shared autonomous vehicles: A review considering car sharing and autonomous vehicles. *Asian Transport Studies*, 5(1):47–63, 2018.

174 Nikhil Menon, Natalia Barbour, Yu Zhang, Abdul Rawoof Pinjari, and Fred Mannering. Shared autonomous vehicles and their potential impacts on household vehicle ownership: An exploratory empirical assessment. *International Journal of Sustainable Transportation*, 8318:1–12, 2018.

175 Diana Jorge and Gonçalo Correia. Carsharing systems demand estimation and defined operations: A literature review. *European Journal of Transport and Infrastructure Research*, 13(3):201–220, 2013.

176 Niels Agatz, Alan Erera, Martin Savelsbergh, and Xing Wang. Optimization for dynamic ride-sharing: A review. *European Journal of Operational Research*, 223(2):295–303, 2012.

177 Masabumi Furuhata, Maged Dessouky, Fernando Ordóñez, Marc-Etienne Brunet, Xiaoqing Wang, and Sven Koenig. Ridesharing: The state-of-the-art and future directions. *Transportation Research Part B: Methodological*, 57:28–46, 2013.

178 Georg Brandstätter, Claudio Gambella, Markus Leitner, Enrico Malaguti, Filippo Masini, Jakob Puchinger, Mario Ruthmair, and Daniele Vigo. Overview of optimization problems in electric car-sharing system design and management. In H. Dawid, editor, *Dynamic Perspectives on Managerial Decision Making*, pages 441–471. Springer, Cham, 2016.

179 Patrícia S. Lavieri, Venu M. Garikapati, Chandra R. Bhat, Ram M. Pendyala, Sebastian Astroza, and Felipe F. Dias. Modeling individual preferences for ownership and sharing of autonomous vehicle technologies. *Transportation Research Record: Journal of the Transportation Research Board*, 2665:1–10, 2017.

180 Peraphan Jittrapirom, Valeria Caiati, A.-M. Feneri, Shima Ebrahimighareh-baghi, María J. Alonso González, and Jishnu Narayan. Mobility as a service: A critical review of definitions, assessments of schemes, and key challenges. *Urban Planning*, 2(2):13–25, 2017.

181 Roni Utriainen and Markus Pöllänen. Review on mobility as a service in scientific publications. *Research in Transportation Business & Management*, 27:15–23, 2018.

182 Jackeline Rios-Torres and Andreas A. Malikopoulos. Impact of partial penetrations of connected and automated vehicles on fuel consumption and traffic flow. *IEEE Transactions on Intelligent Vehicles*, 3(4):453–462, 2018.

183 Liuhui Zhao, Andreas A. Malikopoulos, and Jackeline Rios-Torres. Optimal control of connected and automated vehicles at roundabouts: An investigation in a mixed-traffic environment. In *15th IFAC Symposium on Control in Transportation Systems*, pages 73–78, 2018.

184 Z. Wadud, D. MacKenzie, and P. Leiby. Help or hindrance? The travel, energy and carbon impacts of highly automated vehicles. *Transportation Research Part A: Policy and Practice*, 86:1–18, 2016.

Part II

Energy and Production

5

Control of Electric Power Conversion Systems

Peter Hokayem[1], Pieder Joerg[1], Silvia Mastellone[2], and Mario Schweizer[3]

[1]*ABB Motion, ABB Switzerland Ltd, Turgi, Switzerland*
[2]*Institute for Electric Power Systems, University of Applied Science Northwestern Switzerland, Windisch, Switzerland*
[3]*ABB Corporate Research, ABB Switzerland Ltd, Baden-Dättwil, Switzerland*

5.1 Introduction

Energy sectors and industrial installations have seen a dramatic increase in productivity and efficiency over the last decade due to increased utilization of power electronic (PE) converters to drive their processes. This ubiquity of power converters has radically modified the type and complexity of the underlying dynamics of power grids, power generation, and industrial processes. Conventionally, power grids have been characterized by high inertia elements like synchronous generators (SGs) and induction machines, as well as passive equipment like transformers and resistive-inductive loads. Moreover, stable frequency-dependent loads and generation profiles with a fixed configuration resulted in a predictable and stable grid behavior. In contrast, modern power grids are characterized by the pervasive presence of power electronic converters, whose fast dynamics reduce the intrinsic system inertia and, consequently, the underlying stability margins. Moreover, the shift toward renewable and distributed generation has introduced high levels of fluctuation in power generation.

Industrial loads play a major role in the stability of the overall system as their power levels can go above hundreds of Megawatts per load for large plant installations. These loads are affected by the grid strength experienced at their corresponding points of common coupling (PCC) and in turn can introduce large variations on the instantaneous power demand. However, there are enormous opportunities when operation of these loads can be smartly orchestrated and leveraged to introduce additional distributed stability elements. For example, some loads may have reactive power support or active power adaptation capabilities

The Impact of Automatic Control Research on Industrial Innovation: Enabling a Sustainable Future,
First Edition. Edited by Silvia Mastellone and Alex van Delft.

that have been so far unutilized as they would require additional levels of dispatch controls. The foregoing factors, accompanied by new requirements from grid and industrial processes, give rise to a higher degree of complexity in operating and maintaining all parts of the energy chain: generation, transmission/distribution, and industrial loads.

The challenges associated with this fundamental shift in the nature of power grids have been extensively addressed in the literature [1–3]; however, less research effort has been dedicated to address large industrial loads with complex processes and their interaction with the energy network. Moreover, many of the results presented in the literature may not consider all aspects of realistic scenarios and often offer solutions with a high degree of computational complexity, thus failing to reach the required maturity level to be implemented on a product or a service.

In this chapter, we offer an industrial perspective on the challenges and opportunities posed by the pervasive use of power electronic converters in power generation and industrial processes, with special focus on industrial-type medium-voltage power electronic converters. We present some key directions for automatic control professionals in which they can develop fundamental research with high practical impact that addresses the growing complexity in power conversion systems and helps meet the stringent requirements on performance, stability, and efficiency. The objective is to facilitate a positive shift towards optimal, reliable, and sustainable operation of power electronic conversion systems.

5.2 Power Electronic Conversion Systems

As of 2022, industrial power electronic drives are present in nearly all energy, industry and mobility sectors, with an estimated market size of more than 23 billion USD per year (see Figure 5.1). Some specific applications are:

Mobility applications: marine propulsion, railway systems, E-buses, E-trucks.
Industrial applications: mining, chemicals processing, minerals extraction, oil and gas production.
Power applications: classical generation, wind conversion, hydropower, photovoltaic systems.
Special applications: test stands, wind tunnels.

The massive introduction of power electronic (PE) converters in those applications witnessed in the last decade led to a fundamental shift from minimally interactive elements characterized by high inertia and slow dynamics to highly dynamic and interactive systems. This shift has enabled a wealth of benefits including fast dynamics and high levels of controllability of frequency and power,

Figure 5.1 Examples of various applications of drive systems: power generation, industrial applications, mobility, and smart homes. Source and copyright: ABB.

leading to optimized process performance and energy efficiency. Additionally, power electronic conversion has enabled new dimensions, including renewable generation, compact energy supply and storage, DC/DC power conversion, improved energy reliability with UPS systems, and it has driven the overall trend in electrification.

On the load side, we have seen a shift from constant to variable power loads. The motors driving the processes have been interfaced to the grid via power converters to gain flexibility and control of the variable speed. This results in a more modular, dynamic, flexible, and versatile energy system better and more efficient industrial processes. On the other hand, the fast dynamics are accompanied by reduced stability margins for the overall system, and at the same time, power generation has become more distributed, intermittent, and unpredictable. The emerging new system comes with increased complexity but also potential for overall performance improvements. This evolution led to the formulation of new, at time conflicting, operational requirements for the grid and for large loads to guarantee system stability and simultaneously fully utilize the new potential.

5.2.1 Fundamental Elements and Mathematical Models

Power conversion systems, comprising a power converter, filters, interconnected power sources, storage elements, and loads, enable high-level power conditioning and adaptation to process needs at good energy efficiency levels over a wide range of power demands. Specifically, these systems are responsible for transforming electrical energy to mechanical energy in industrial applications and mechanical energy to electrical energy in power generation, see Figure 5.2.

Driving

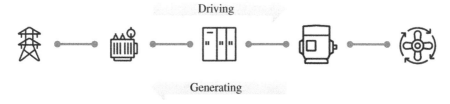

Generating

Figure 5.2 General power converter and bi-directional power flow. Left to right: grid, transformer, power converter, motor/generator, process/driveline. Source and copyright: ABB.

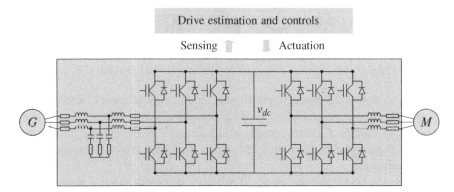

Figure 5.3 General two-level converter connected to an ideal grid (G) and a motor (M).

Additionally, electric power is converted between various voltage levels and forms (AC/DC). For example, utility systems usually generate, transmit, and distribute power at a fixed frequency (50 or 60 Hz) while a consumer may require power in AC or DC form, with variable frequency correlated with the process speed. Power electronic converters are the core of power conversion systems. These converters are based on power semiconductor devices of various power handling capabilities and switching rates ranging from several tens of Hz to a few tens of kHz. Classically, semiconductor devices have been silicon-based. More recently, switching devices based on wide-bandgap (WBG) materials such as silicon carbide (SiC) or gallium nitride (GaN) have become more common. Converters with WBG devices allow more efficient transformation and control of electrical power, with reduced device size and weight. Typical efficiency levels exceed 95% and reach up to 99% [4].

To control the electric power, PE converters control the voltages and currents on both the grid and the load sides based on specific process needs. Controllers typically utilize averaging methods, in which the switching dynamics are neglected due to their high-frequency content, and continuous control inputs are assumed instead for controller design. In what follows, we refer to Figure 5.3 and provide simplified models of the various system components. See, for example, [4–6].

On the grid side, the simplified dynamics are modelled as[1]

$$\frac{d}{dt}I_{grid} = -\frac{R_{grid}}{L_{grid}}I_{grid} + \frac{1}{L_{grid}}(V_{grid} - V_{rec}) \tag{5.1}$$

where I_{grid}, V_{grid}, R_{grid}, and L_{grid} are the grid currents, voltages, resistance, and inductance, respectively, and V_{rec} is the voltage provided by the converter. The

---

1 We neglect passive filtering effects and simplify the notation.

voltage can in turn be expressed in terms of the switch positions σ_{rec}, i.e. the control inputs, and the DC capacitor voltage as

$$V_{rec} = \sigma_{rec} v_{dc} \tag{5.2}$$

The capacitor voltage v_{dc} dynamics are affected by the grid currents, the motor currents and the switch positions on the rectifier and inverter side as

$$\frac{d}{dt} v_{dc} = \frac{1}{C} \left(\sigma_{rec}^T I_{grid} - \sigma_{inv}^T I_{motor} \right) \tag{5.3}$$

Under high switching frequency assumptions, both σ_{rec} and σ_{inv} are approximated as continuous functions rather than taking discrete values. On the inverter side, the dynamics of an induction motor can be expressed as

$$\frac{d}{dt} \psi_s = -R_s I_s + V_s$$

$$\frac{d}{dt} I_s = M_1 I_s + M_2 \psi_s + M_3 V_s$$

$$T_{elec} = \frac{3p}{2} \psi_s \times I_s$$

$$J \frac{d}{dt} \omega_{mech} = T_{elec} - T_{load}$$

where ψ_s, I_s, and $V_s = V_s(\sigma_{inv}, v_{dc})$ are the stator flux, current, and voltage vectors, respectively, $M_1 = M_1(\omega_{mech})$, $M_2 = M_2(\omega_{mech})$ and M_3 are matrices, some of which depend on the motor parameters and the mechanical speed ω_{mech}, and T_{elec} is the generated electric (air-gap) torque that regulates the mechanical speed ω_{mech} and maintains it at some desired reference, given a certain load torque T_{load}, and R_s and p are the stator resistance and the number of electric pole-pairs, respectively. The resulting model of a single PE conversion system is nonlinear with a discrete set of inputs. In an industrial application, one typically encounters several inter-connected PE conversion systems with various driven processes, see Figure 5.4, comprising several stages of mechanical elements (shafts, couplings, gearboxes, compressors, etc.). Accordingly, the dimension of the state-space grows and con-trol design becomes extremely complex to resolve in a centralized way. As such, control design has been mainly restricted to a single power conversion system with simplifications in order to render the control problem tractable. These sim-plifications may, however, introduce instabilities due to unmodelled dynamical interactions, which may lead to undesired system shutdowns or even failures.

5.2.2 Operation of Power Converters

Power converters are generally operated according to the control structure depicted in Figure 5.5. Both grid- and motor-side power converters share the same

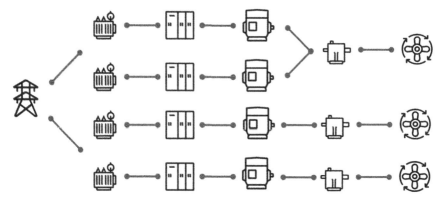

Figure 5.4 Complex multi-drive industrial setup with several rotating equipment applications. Source and copyright: ABB.

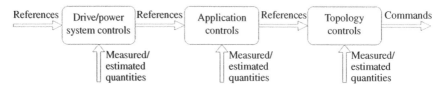

Figure 5.5 Abstraction of the various control layers engaged in a drive system control.

structure of cascaded control loops operating over different frequency ranges to fulfill the separation principle in the frequency domain.

5.2.2.1 Topology Controls

The fastest control layer, sometimes referred to as the *topology controller*, is agnostic to the control objective of a specific converter unit. It operates in a dynamic range between a hundred to a few hundred Hz and 10 kHz and controls the flow of electric energy by applying suitable switching commands (σ_{rec}, σ_{inv}) to the semiconductor switches, thus providing the required power transfer. One part of the topology controller orchestrates the switches and can use switching patterns based on PWM (pulse-width modulation) or OPP (optimized pulse pattern) schemes [5, 7]. Due to the high switching frequency of the underlying semiconductor components of the converter, it is possible to design a second part of the topology controller to regulate the current without considering the specific converter topology (Figure 5.6).

There are several versions of the topology controller, depending on the applications and requirements. For example, the typical *current controller* on the rectifier side provides an AC voltage reference to the inner control loop with

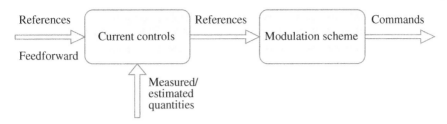

Figure 5.6 Topology controller: example of current control and modulation scheme.

a bandwidth in the range of 10 s to few hundred Hertz and receives its target reference values from the outer loop controller based on the required power to be delivered. Another example is the current controller on the motor or inverter side, which receives its reference based on the required electric torque T_{elec} to be delivered.

A different variant of the topology control on the motor side is direct torque control (DTC) [8], in which the torque and flux magnitude references coming from the application controller are directly mapped to voltage switch positions without any intermediate generation of a voltage reference to be "modulated."

5.2.2.2 Application Controls
The next level of control is often referred to as the *application controller*, which is responsible for the main application objectives of the inverter or rectifier unit and operates with a bandwidth of 10 Hz to few 100 Hz. For example, control objectives may be to track a given active and reactive power reference for a grid-connected converter, to maintain the capacitor DC voltage at a prespecified level, or set the speed of a motor at a given reference for a drive. Variants of this approach to control the rectifier look directly at the power variables (active and reactive) and provide active and reactive current references to the inner control loop [9, 10].

5.2.2.3 System Controls
The outermost loop is the *drive or power system controller*, which typically operates in the few Hz range and orchestrates the references given to the individual drives. For example, this could be a power-level dispatching layer in a wind farm or a speed or frequency droop controller that coordinates the actions of several electric drives [9].

5.2.2.4 Estimation
In addition to the aforementioned control layers, the operation of power converters relies on state estimation elements that provide the controllers with information about quantities that cannot be directly measured. For example, on the grid side,

a grid synchronization phase-locked loop (PLL) running at 10–100 Hz enables to separately control active power and reactive power. PLLs are usually reliable in the case of balanced grids; however, they are sensitive to phase jump and weak grids. Moreover, PLL interaction with the control layers poses a challenge to the overall system stability [10]. On the machine or process side, typically machine flux and speed estimators are implemented. Note that the effectiveness and speed of estimation would influence the stability of the overall system.

We will revisit some of the above control levels in the upcoming sections and provide some more details whenever needed.

The state-of-the-art operation of power converters follows the aforementioned control structure and is based on standards defined for classic scenarios of electrical grid and processes characterized by fixed environmental conditions, high stability margins and limited requirements. One main limitation of the classical control structure is the independent design of interactive control loops and protection functionalities, specifically calibrated for ideal or nominal operating conditions that lead, at best, to reduced system performance and in the worst-case to system instability. Addressing the new challenges and requirements via the optimal configuration and operation of the power converters presents a great opportunity to maximize the benefit of the modern power conversion system by improving efficiency, reliability and safety, and reducing investment and operational costs over the complete system life-cycle. In the following, we will focus on the operation of the back-to-back converter that includes the AC to DC and DC to AC conversion stages. We will consider power converters operating in two scenarios:

Grid-connected converters: A grid characterized as an open unknown system where converters operate mostly as interfaces to power generation elements, with power flowing from the generation into the grid. In this scenario, the converter has no knowledge of the grid and sees it as a simplified Thevenin equivalent model for the purpose of control design.

Industrial drives: Industrial processes as an isolated network of variable-frequency loads fed by the grid. In this scenario, the power flows from the grid to the loads and the process requirements include electromechanical aspects. In this scenario, the converter has more knowledge of its interface with respect to the grid scenario, and some of the mechanical dynamics are known of the equipment being chosen and installed. Modeling based on first principles is possible and then some abstraction for the control design can be used at a second stage.

This classification based on power flow direction includes some exceptions such as loads operating in regenerative modes and producing power to be stored in batteries or injected into the grid. Moreover, it is foreseeable in the future that loads will be required to provide grid support and ancillary services. In those cases, the loads will operate as generation units.

In what follows, we highlight for each of the two scenarios the challenges and propose research directions to pursue. The proposed paths present opportunities that will enable the next generation of energy and production systems to reach their highest potential in maximizing productivity and energy efficiency in a sustainable way.

5.3 Grid-Connected Power Converters

We now consider grid-connected voltage source converters (VSCs) regulating the power exchange between the grid and power generation, storage, or consumption elements. In the electrical grid, the growing presence of power electronic converters has radically modified its dynamics, introducing reduced stability margins and unpredictability. Following this shift, additional requirements for the converters have been introduced in the form of grid codes. In this section, after introducing the grid standard requirements and the converter control structure, we detail upcoming requirements and specific challenges and finally offer a perspective on possible research directions to enable the optimal operation of grid-connected power converters geared toward system stability, reliability, and delivery of high-quality power.

5.3.1 Standard Requirements and Control Structures

Traditionally, the grid has been dominated by grid-feeding power converters operating as current sources to inject active power into the grid (see Figure 5.7), as described in the current IEEE Standard 1547 [11, 12], which requires unity power factor at the output of an inverter. An extended version of this standard, 1547.8 [13], proposes voltage-based reactive power control as a methodology for adapting the reactive power generation of an inverter to assist the power quality. In particular, the grid voltage stabilization and supply of reactive power is typically regulated by a droop control at a rate of approximately 10 Hz. According to the standard, an inverter monitors its terminal voltage and sets its reactive power generation based on a predefined Volt-Var droop curve.

In addition to the international standards, region-specific grid codes define frequency, voltage, and power rating in the steady-state for generation assets. Typically, continuous operation is required around the rated frequency of the grid. During transient operation, and for some limited time period, generators are allowed to operate with frequency values outside the allowed band.

Voltage standards define the maximum allowed deviation from the nominal value for both service voltage, measured at the point of delivery, and utilization voltage, measured at the terminals of the utilizing equipment. This indirectly defines admissible voltage drop across the lines. The voltage security levels vary in distribution and transmission grids and across regions. Finally, more advanced requirements include power quality, e.g. harmonic regulation, and protection functionalities. Besides the external grid requirements, the converter control needs to meet internal requirements to guarantee safe and reliable operation in presence of external disturbance and security constraints. The grid codes and inverter requirements determine the cascaded control structure, including the topology controller and the application controller that regulates the active and reactive power or, similarly, the voltage at the PCC.

Depending on the energy source, specific application controllers provide references for the inner loop topology controllers of the converters. Below we list some examples:

Synchronous generators (SGs): Application controllers include an automatic voltage regulator (AVR), a power system stabilizer (PSS) for frequency stabilization, and a governor or prime mover control.

Photovoltaic (PV): Application controllers include a maximum power point tracker (MPPT), DC link voltage control; the topology controller includes a current controller and various modulation schemes.

Wind turbines: Application controllers include an MPPT, DC link voltage control; the topology controller includes a current controller and various modulation schemes.

Energy storage systems (ESSs): Application controller includes the DC voltage control; the topology controller includes a current controller and various modulation schemes.

Fuel cells: Application controller regulates the DC power (DC/AC).

5.3.2 Prospective System Requirements

It is expected that future grids will be populated by grid-supporting and grid-forming power converters, current or voltage-source based, that will provide frequency regulation and inertia support to the grid [1], see Figure 5.7. With increasing interconnectivity levels of heterogeneous elements in the grid, interoperability requirements are also expected to change. Interoperability standards, including on harmonics injection, are currently in place in some countries for specific applications and are expected to be further extended in many fields [12].

One concrete example of upcoming new standards for electrical grids can be found in the UK, where in a 2022 report [14], the UK national electricity system operator (National Grid ESO) addresses the urgent need for measures to ensure grid stability in the UK as synchronous generator-based power generation is partially or fully abandoned. The main idea is to develop the minimum Grid

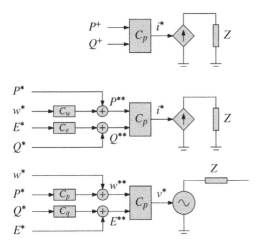

Figure 5.7 Control strategies for grid-connected, grid-supporting, and grid-forming power converters.

Code with technical specification for a grid-forming capability. The market arrangements will then be addressed separately once the specification and technical requirements are in place. The report presented by the NG ESO is a good example of the new ancillary services that are expected to become a requirement in future grids. The services are exclusively derived from the natural behavior of the synchronous generator, namely: active control-based power, active phase jump power, active damping power, active inertia power, active rate of change of frequency (RoCoF) response power, control-based reactive power, fast fault current injection, grid-forming active power, grid-forming capability, and voltage jump reactive power.

The report also describes the minimum requirements on data exchange between grid operators and potential suppliers of the new ancillary services.

Another example, specific for interoperability standards, has been defined and is currently adopted in some counties for railways networks. Specifically, the interoperability standard defined for the Swiss railway system requires each PE interfaced locomotive connecting to the railways grid to be certified to meet specific impedance requirements [15]. This came as a result of a grid collapse in 2005, when the sudden introduction of power electronic converters led to an excess of harmonics content in the grid, with consequent overvoltage and collapse. With higher penetration of PE converters, the same phenomena can be expected in other grids, including distribution grids, microgrids, wind and PV parks and industrial grids, which will lead to the introduction of additional interoperability standards for those grids as well. With new interoperability requirements in place, two major technical and research challenges will be (i) how to control the power converter so to guarantee that the standards are met, and (ii) how to test and certify a controller for those standards.

5.3.3 Control Challenges

In the past, the power grid was dominated by rotating machinery, i.e. generators and electrical machines directly connected to the grid. Consequently, the dynamics of the grid were determined by the physical behavior of these machines and other passive elements such as transformers, cables, transmission lines and capacitor banks. Converter-interfaced resources only made up a small fraction of total sources/loads (<10%) and their behavior could mostly be neglected. Analysis, planning, and operation of the power grid were based on tools and schemes historically grown and heavily optimized for the described scenarios. Standards or grid codes regulating the connection and behavior of converter-interfaced sources and loads were sketched with the assumption that their impact on the overall behavior of the power grid is marginal. Conventional control schemes for converters assume that the power grid is an infinite resource (infinite bus, see Figure 5.8) which can deliver/absorb active and reactive power without any reaction. Today we face a different situation (see Figure 5.9), which will be even more pronounced in the near future. Massive integration of renewable energy sources (RESs) and accelerated electrification of transport and industry have led to the situation where at certain hours in the year, most resources are connected with PE converters to the grid. Especially in smaller power grids, e.g. in Ireland, this situation has become very common. Consequently, the dynamics and behavior

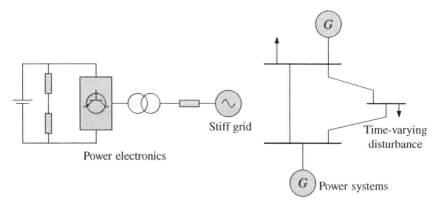

Figure 5.8 Representation of the electrical grid as a stiff element as seen from a power converter and of the power converter as a varying disturbance as seen from the grid.

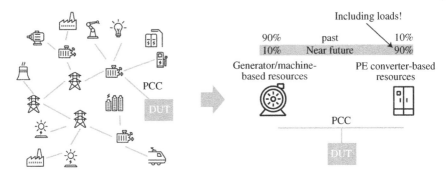

Figure 5.9 The past and future energy systems. Source and copyright: ABB.

of the power grid are dominated by the behavior of these converters, which in turn is mostly determined by their control schemes. However, some rotating machines still remain in the grid and their physical behavior must be respected and cannot arbitrarily be overruled by new control behavior. Furthermore, fault handling and protection mechanisms of the power grids are based on the interplay with fault behavior of synchronous machines. In a way, modern control solutions for the converter-dominated power grids have to cope with the legacy equipment and need to be compatible to a certain extent with historically grown operation schemes of the power grid. Currently, the majority of grid-connected PE converters is controlled with a standard cascaded loop scheme, described previously in Section 5.2. With this control configuration, the converter behaves toward the grid as a constant power sink/source in the low-frequency range ($f < 150\,\mathrm{Hz}$). In the high-frequency range ($f > 150\,\mathrm{Hz}$), the equivalent terminal behavior can be described by a current source that is connected in parallel with a rather high-output impedance ($Z = 1 \ldots 10\,\mathrm{p.u.}$), i.e. the converter current does not react to changes in the grid voltage. Existing standards [16] limit the current harmonics that can be injected into the grid, i.e. the line currents need to be as sinusoidal as possible. Rotating machines behave quite differently. All rotating machines induce a so-called back-emf at the nominal operating point, which is a voltage source. The model is augmented with a series-connected impedance of low value ($Z = 0.1 \ldots 0.5\,\mathrm{p.u.}$) representing the inductive nature of the stator winding. Although this model is oversimplified and does not capture accurate behavior in many abnormal operating conditions and fault cases, it represents well the behavior toward the grid under normal conditions. Most of the challenges in the modern power grid can be understood with a simple aggregate model that combines all rotating machines and represents them with a single equivalent voltage source and series impedance, and all converters are aggregated and represented as a current source with parallel impedance, see Figure 5.10. However, to capture also low-frequency effects related to mechanical rotor speed

Generators behave as a voltage source
$$Z = 0.1...0.5 \text{ p.u.}$$

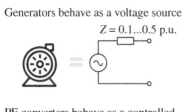

PE converters behave as a controlled current source

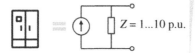

Aggregate model of all direct online (DOL) connected machines and generators

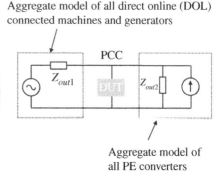

Aggregate model of all PE converters

Figure 5.10 Aggregate models of DOL machines and generators, and of PE converters. Source and copyright: ABB.

oscillations (inertia and damping power), the nominal frequency of the voltage source is not fixed but can vary in a limited frequency range depending on the overall dynamic loading conditions.

If the share of rotating machines is reduced from 90% to 10% and in turn, the converter-interfaced resources are increased from 10% to 90%, some fundamental changes can be observed at arbitrary connection points to the grid. Several scenarios are also discussed in [17–21].

5.3.3.1 Lack of Inertia and Power-Damping Capabilities

Unplanned major changes of the overall loading condition in the grid happen regularly and are related to unforeseen faults and tripping events, e.g. tripping of HVDC lines or large generator assets, or in the worst case, a dangerous split of the synchronous area due to loss of key transmission lines. How will the power-generating units react to the changed loading condition? The underlying power sharing problem can be represented with the simplified block diagram in Figure 5.11 where two voltage sources with inner output impedances are connected in parallel and have to supply a step change in power consumption. In the traditional power grid, the two voltage sources would represent two aggregated groups of SGs. After connection of the load, which represents the changed loading condition due to the contingency event, the power redistributes within milliseconds according to the terminal characteristics of the aggregated assets, see Figure 5.12. At the very beginning (0–20 ms), the power splits according to the ratio of the output impedances. In a subsequent phase, the rotor frequencies of the machines start to change because the changed output power is fed from

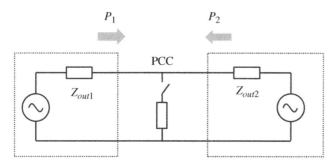

Figure 5.11 Simplified aggregated model with two voltage sources to illustrate the sharing of power transients between power generating units.

the energy stored in the rotating mass, represented by the inertia. A decay and a low-frequency oscillation of the rotor frequency is triggered, damped by the effect of induction machines and damper windings on the SG. Finally, prime movers are actuated according to the droop control feedback law and increase/reduce the mechanical input power to the machine until the frequency stabilizes at a slightly lower/higher level. Top-level control would be activated at a much slower timescale to bring back the frequency to its nominal level.

Imagine now the described change from 90% to 10% SG and 10% to 90% converter assets. All rotating machines are aggregated and represented by the

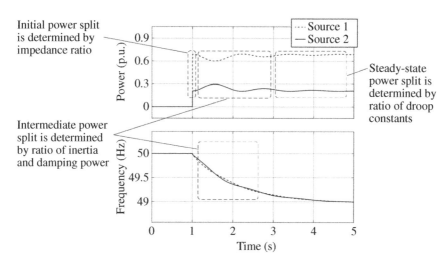

Figure 5.12 Typical evolution of power and grid frequency as a reaction to a step change in loading condition.

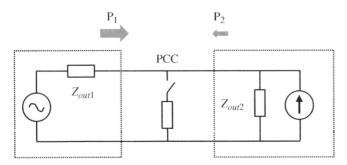

Figure 5.13 Simplified aggregate model with a voltage source representing all synchronous generators and a current source representing all PE converters.

voltage source model and all converters are aggregated and represented by the current source model, as depicted in Figure 5.13. Again, a large load is connected/disconnected at the PCC. The total equivalent grid impedance (parallel connection of Z_{out1} and Z_{out2}) will increase; the total inertia, damping power, and droop feedback will reduce significantly. The same power step will lead to an immediate increased dip in voltage amplitude because of increased equivalent grid impedance. In the described cascade of events, with the exception of the very first instant determined by the ratio of output impedances, the conventionally controlled PE converter assets do not intervene, i.e. their injected current or power (dashed curve in Figure 5.14) is unchanged. The transient and steady-state change in power will be supplied uniquely from the remaining SG, resulting in an overload. The frequency excursion will increase due to reduced inertia, damping power, and droop feedback power.

As a first improvement step, droop control has been introduced in many converters several years ago (BESS, wind, PV inverter downregulation), i.e. the consumed/injected power is changed proportionally to the measured grid frequency. However, this has only partly resolved the issue, as can be seen in Figure 5.15.

With droop control extension, the converter-interfaced assets contribute to the power sharing and reduce the frequency excursion in the steady-state. However, in the very first moment, the remaining SGs are still being overloaded by large transient power steps, as the converter-interfaced assets do not contribute to the power-sharing sufficiently during the transient phase. This leads to high df/dt and increase wear-and-tear to rotating machinery with early material fatigue. The issue is related to the rather high output impedance of the controlled converters and missing inertial/damping power injection. Further changes to the traditional converter control schemes, such as virtual synchronous generator (VSG) control [22–24], can relax the remaining unbalanced power sharing during a change in

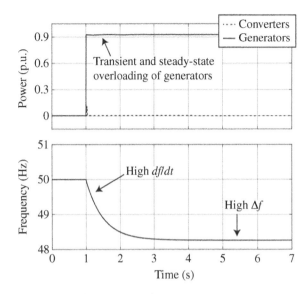

Figure 5.14 Power sharing between generators and converters and grid frequency excursion as a reaction to a significant change in loading condition in the power grid in the case where converters are controlled in a conventional way without any grid-supporting feature.

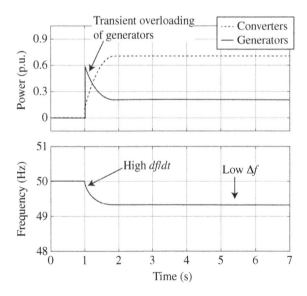

Figure 5.15 Power sharing between generators and converters and grid frequency excursion as a reaction to a significant change in loading condition in the power grid in the case where converters are controlled with simple steady-state droop control.

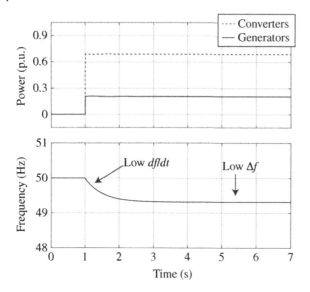

Figure 5.16 Power sharing between generators and converters and grid frequency excursion as a reaction to a significant change in loading condition in the power grid in the case where converters are controlled with grid-forming control, e.g. virtual generator control.

loading condition. VSG control differs fundamentally from the conventional converter control scheme and has not found widespread application in commercial converters yet, with the exception of microgrid applications. VSG emulates the terminal characteristics of a synchronous machine, i.e. a low output impedance combined with inertial and damping power response. Converters equipped with VSG control or an equivalent control scheme would participate in power sharing on all time scales, as shown in Figure 5.16. Overloading of remaining SGs is avoided and the grid frequency excursion, as well as the df/dt, is minimized. However, the know-how about VSG control in industry is limited and the method still presents partially unsolved challenges:

- Proper tuning of the VSG control to avoid oscillations and improve damping.
- Difficult behavior during faults on the grid side. There is a need to strictly limit the current in converters to avoid damage to the semiconductors. However, classic VSG control does not directly control the currents and consequently leads to high overcurrent during faults. Nested current limitation schemes and switchover logics are often required, which are difficult to tune and can lead to resynchronization issues after fault clearing.
- Contradictory requirements of the process control and adverse interactions (e.g. DC-link voltage control, control of constant load power, and maximum power

point tracking). As the injected power is partly determined by the state of the power grid, it is difficult to meet superior process-related control goals.

- Converter design and economic impact. Operation with headroom might be required to emulate the inertial response, which clashes with maximum power point tracking.

5.3.3.2 Harmonic Power Sinking/Supplying

Another issue related to standard converter control is the sinking/supplying of current harmonics. Today, many heavy loads are interfaced with diode or thyristor rectifiers. These rectifiers inject current harmonics into the grid that typically flow to the SG with low output impedance. Imagine again the tenfold reduction of online SGs and the related increase in equivalent grid impedance. The same current harmonics will lead to much higher distortion of the voltage waveform at the PCC and reduce the voltage quality substantially, see Figure 5.17. In the worst case, sensitive equipment connected to the PCC will stop operating. Again, this is related to the high output impedance of current-controlled converters and the standards enforcing sinusoidal currents from converter-interfaced assets. VSG control emulates not only the inertial and damping power of a synchronous generator but also its low output impedance. Consequently, it is an adequate solution to improve harmonic power sharing in the converter-dominated grid. However, also for this scenario there are several unsolved challenges:

- If the VSG is implemented with cascaded voltage and current control schemes to improve the overcurrent protection, it is difficult to achieve sufficiently high bandwidth of the current control loop to accurately emulate the low output impedance in the higher frequency range.
- Low output impedance leads to the flow of harmonic currents, which contradicts certain grid codes. This is more of a regulatory challenge.
- The flow of harmonic currents imposes design challenges and additional losses in many components, such as filter inductors, coupling transformers

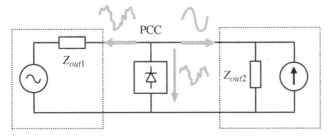

Figure 5.17 Reduced bus voltage quality because of increased equivalent grid impedance.

and capacitors. Sufficient margin for harmonic current capability on top of the fundamental current is required for the converter and may lead to an increase in converter costs.

• Harmonic currents may jeopardize the overall control performance because of difficulties related to sampling and separation of fundamental and harmonic components.

5.3.3.3 Supplying Asymmetric Loads

Similar issues are observed in grids characterized by asymmetries due to asymmetric loads. In that case, negative sequence currents are required to flow. These currents are typically sourced from the remaining SGs. With reduction of the remaining online SGs, similar asymmetric loading conditions will lead to an increased voltage imbalance at the PCC. Conventional current-controlled converters inject symmetric three-phase currents and consequently do not counteract the resulting voltage asymmetry. Again, VSG control can resolve the issue because the emulated voltage source behind a low output impedance allows asymmetric currents to flow. However, this behavior presents once more the following challenges:

• If the VSG is implemented with cascaded voltage and current control schemes to improve the overcurrent protection, it is difficult to achieve sufficiently high bandwidth of the current control loop to accurately inject the required negative sequence current.

• Negative sequence currents give rise to power pulsations at the DC side of the converter. This translates to design changes, e.g. increased DC-link capacitance with related losses and costs.

5.3.3.4 Increased Risk for Electrical Instability

Another related issue that is often overlooked is the effect of unknown resonances in the interconnected systems. Next to the electromechanical instabilities related to oscillations in a frequency range below 10 Hz, which is assessed traditionally in grid stability studies, there are also resonances at higher frequencies (200 Hz–2 kHz) in a grid with low damping. These resonances are related to capacitor banks, transformer inductances, converter input and output filters and converter control behavior, see Figure 5.18. With the described change from 90% to 10% SG and 10% to 90% converter assets, the equivalent grid impedance seen at the PCC tends to increase. Electrical resonances with capacitor banks shift to lower frequencies, where the damping is typically reduced because of reduced harmonic losses. Furthermore, converter control may lead to frequency regions where the converter output impedance is negative, e.g. due to constant power load behavior at low frequencies, or due to limited control bandwidth and delays

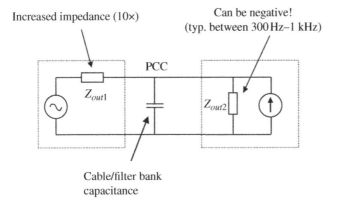

Increased impedance (10×)

Can be negative!
(typ. between 300 Hz–1 kHz)

PCC

Z_{out1}

Z_{out2}

Cable/filter bank
capacitance

Figure 5.18 Weakly damped resonance with capacitive filter bank due to interaction with converter control.

at higher frequencies. Such negative impedances can further reduce the damping of electrical resonances in the grid or even lead to system instability.

5.4 Industrial Drives

As of 2022, electric motors consume over 45% of the world's electricity and by 2040 the number of motors is expected to double. Depending on the industry sector, between 10% and 90% of all motors are controlled by power converters, allowing higher levels of energy efficiency over a wide range of power and process demands.

5.4.1 Standard Requirements and Control Structure

Traditionally, processes have been driven with a fixed speed and directly connected to the grid. The introduction of power converters has removed this restriction and enabled a shift toward potentially highly dynamic processes that are, however, not operated at their full potential. As mentioned in Section 5.2.1, actuation capabilities inside the converter are of a discrete nature, i.e. the output voltage levels can only be generated in steps. This gives rise to electrical harmonics, which propagate through the system and are fed back to the grid or to the motor side. A second cause of harmonic distortion is the limited switching rate, typically bounded to minimize the switching losses in the high-power converters – a major part of the overall losses of the drive – and directly related to the switching frequency. The limited switching frequency and the presence of low-order harmonics limit the capability of the controller that operates under the assumption of an averaged model where the voltage at the inputs and outputs of an electric drive are purely sinusoidal. To compensate for these shortcomings, additional requirements on the harmonics content are imposed on the converter controller by the grid codes [16]. On the machine side, the harmonics content is characterized by the total harmonic distortion (THD) of the current. The THD is a measure of the collective strength of all the higher-order harmonics compared with the fundamental component, i.e. the undesired content besides the fundamental component. Additionally, on the process side, there are robustness requirements such as accurate tracking of the speed reference in the presence of disturbances. This is typically expressed in terms of acceptable speed variation following changes in the unknown process torque, and it ensures high quality of materials production, for example, in metals applications. For complex mechanical setups, there are the API 617 and API 684 standards [25, 26] that provide guidelines to design and analyze stability, lateral and torsional, for complex rotating drivelines, see Figure 5.4.

While in the past it was acceptable to "trip" or stop operating the converter in case of minimal voltage or frequency deviation, today, increased resilience and partial availability are required to operate the system at its performance limit during transient conditions and avoid unnecessary shutdown. Moreover, due to the large interconnectivity characterizing industrial drive systems, several new standards focus on interoperability and harmonics requirements that concern the whole system. Interoperability is broadly defined as the capability of two or more components, devices, systems or applications to exchange and use data securely. This property enables utilities to integrate multiple different technologies with the assurance that they will work together and mitigate the risk of instability. Interoperability is stated as a principal enabler of the new system control schemes necessary to manage the active participation of distributed resources. One specific example of interoperability requirements for interconnected systems, i.e. loads, converters and generators, are harmonics restrictions to limit the harmonic distortions in the electrical and electromechanical systems.

The power converter control architecture is designed to meet the described operational requirements. Referring to Figure 5.5, the control structure of an industrial drive has two application layers, one for the grid side and one for the process side; the former caters for maintaining the internal DC voltage of the decoupling capacitors at some given level and the latter caters for tracking of a required process speed set-point. Additional feed-forward terms ensure a fast response during transient conditions. Moreover, the application controllers include protection mechanisms, for example, limiting the motor speed at some given level, controlling the rate of change, riding through disturbances, or performing emergency stop ramps. Underneath the two application controllers, there are the topology controllers mentioned earlier. On the grid side, a current controller and a modulation scheme is used, typically with harmonic emission restrictions as part of the modulation scheme. In most industrial sites, the processes are complex (see Figure 5.4) and are, as such, driven by several motors. Coordinating the application layer of the several motors is achieved by a third layer of automation, the *drive system control*, which performs scheduling tasks and coordinates the various motor drives through sequences required for operation.

5.4.2 Prospective System Requirements

Besides the basic performance requirements described in the previous section, there is a growing trend towards enabling industrial loads to contribute to grid stability and power quality. Here, the same requirements on ancillary service provision that are currently in place for the sources might be extended to industrial loads. Although the operation principle would be the same for sources and loads, new challenges will arise due to a necessary trade-off between the requirement to provide constant power to the process and the new requirement to support the grid. For certain processes and loads that do not have strict requirements on constant power supply, e.g. heating, ventilation and air conditioning, an adaptation of the consumption over time might be admissible. This degree of freedom on the load side can be leveraged in support of the grid side. For example, emulating very fast services such as virtual inertia with loads or adapting the steady-state power consumption to support grid frequency regulation. One example of an upcoming standard in this direction is the "AS 4755 Demand Response Standard" [27] prepared by the grid operator AEMO of the Australian power grid. This standard describes technical requirements for various load classes and applications that could contribute to grid support requirements. Similarly, AEMO is discussing the introduction of new reserve market products for virtual inertia provision that are technology open, i.e. not only sources but also loads can contribute. Another example that extends the control requirements for grid-side renewable generators to grid-side industrial applications is anticipated, for example, by the proposed modifications to the UK grid code GC0137 [14]. The latest edition of the grid code sets the arena for well-defined ancillary services: control-based reactive power, fast fault current injection, and voltage jump reactive power. With some coordination between load and grid, further services could be provided with some economic incentives, such as Active RoCoF response power and active damping power. The grid code anticipates that markets will emerge for these types of services that could technically be provided by a (non-regenerative) motoring variable-speed drive without compromising the motor performance.

5.4.3 Control Challenges

In Section 5.3, we focused on control challenges related to grid-connected converters operating in generation mode, i.e. providing power to the grid. Industrial drives powering a process have been predominantly operating as consumption elements drawing power from the grid. With the new generation of grid codes previously discussed, industrial drives will have to operate flexibly in order to enable power flow in both directions. This poses additional control and operation challenges both on the grid and on the process side.

5.4.3.1 Grid Side

The operation of the grid side of industrial drives faces similar challenges to the ones already described in Section 5.3. The topology controller layer needs to handle the requirements on harmonic emissions and other frequency-domain-specific phenomena, which are becoming more stringent over time. These requirements are tendentially conservative as they consider only the capability of a single topology controller and an assumed worst-case scenario for harmonics emission. Conditions can be relaxed to consider several drives that are connected to the same PCC, as seen in Figure 5.4, thus gaining additional flexibility in managing the overall harmonic emissions at the PCC.

An example of frequency-domain requirements is impedance profile requirements that are currently adopted in railway systems (traction, transmission, and distribution) [15] and are expected to be extended to grid-side inverters of industrial variable-speed drive systems. More elaborate codes will impose a coupled requirement on frequency-domain performance and active control of reactive power consumption for industrial drives. Advanced control methods, integrating time and frequency domains, see for example [28], could optimally balance tracking objectives and constraints on harmonic emission.

5.4.3.2 Application Controls

Currently employed application control layers on the motor or process side apply basic PID controls to achieve speed tracking by issuing torque references to the topology layer. In a scenario where the integrated electric drive, comprising the grid and the motor side inverters, is required to contribute to grid stability and power quality, then the application layer on the motor side cannot only regulate the speed. A multi-objective controller needs to be designed to handle the process and grid requirements according to predefined priorities. Naturally, this gives rise to an optimization-based control scheme that can, in addition, incorporate operational boundary constraints, for example, maximum allowed motor speed, DC voltage levels inside the electric drive, and grid power requirements, to name a few. This scheme can be independent of the grid-side application controller, or preferably, both application layers can be merged into a single controller. First attempts in this direction can be found in [29], in which several modes of operation (steady-state and transient) are optimally combined in a reconfigurable control solution. A more holistic view that would consider a common application control layer for a drive system or systems that orchestrate the power flow, speed control, voltage control, would be of paramount importance for future industrial drives controls.

5.4.3.3 Complex Drivelines

In a typical driveline, an electrical motor supplies torque to a mechanical rotation system to achieve the desired rotational speed. Speed response depends on the motor inertia, additional elastically coupled rotational inertias, and variable loads applied to some of the inertia elements. In the simplest scenarios, the motor is connected through a very stiff shaft to another rotating equipment. However, a mechanical driveline may comprise mechanical components such as a motor, a gearbox, and a compressor that are connected along one or more flexible shafts. In this case, the simplified assumption of a fully rigid system is not valid. Some of the more complex scenarios include multiple electrical machines interconnected to multiple rotating components via a common shaft or via one or more gearboxes, see Figure 5.19. In all cases, the driveline or parts thereof behave as multiple rotational elements with soft coupling, which cause torsional oscillations in the drive system [30, 31]. This oscillation affects the system lifetime and can lead to failures

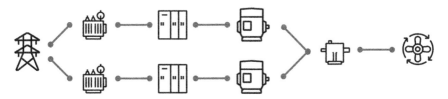

Figure 5.19 Complex driveline with two power converters and two motors connected to a gearbox that drives a process. Source and copyright: ABB.

of certain parts due to fatigue. This constitutes another challenge for the control system to damp the system torsional oscillations. The so-called Active Damping (AD) control methods come into play and complement the passive damping capabilities of the designed driveline [32].

5.5 Research Roadmap

Appreciating the challenges that are faced by power conversion systems, be they of grid-connected or industrial type, is the first step to develop effective decision-making and control solutions for systems of the next generation. There are several research directions in control and decision-making that are necessary to innovate towards sustainability in power conversion applications, and here we outline a few:

Modeling and analysis: In Section 5.2.1, we considered a simplified model of a single electric drive consisting of a nonlinear system with a discrete set of inputs. This does not take into account any complex grid behavior, driven process, or even the multiplicity of interacting power conversion systems. System modeling is a crucial step toward stability and interoperability analysis and control design. Modeling supports in performing transient analysis, understanding the system boundary conditions, and identifying dominant dynamics. Moreover, complex process and driveline models are typically not considered for stability analysis of complex systems and integration of such models during the analysis phase of a complex system would minimize unwanted behaviors due to "hidden dynamics." Furthermore, models resulting from first principles would benefit from available online operational data that would refine the underlying model parameters. This goes in the direction of having online connected digital twins that monitor and learn from an operational system.

Estimation: System identification and state estimation play a fundamental role in extending the information space for control and diagnostics purposes by capturing the converter interaction with other, unknown parts of the systems including other converter units, loads, and the grid. Estimation of the equivalent grid impedance at the PCC can provide important information on the grid behavior and strength that allows the converter to reconfigure and operate optimally. Besides adapting the control modes and/or feedback gains, estimation of the grid impedance is used for protection functionalities, such as islanding detection. On the mechanical side, typically, the system parameters are partially known up to the motor or generator in the best case.

More complex drivelines connected to the drive are typically unknown or their properties, for example, physical damping capabilities, are changing over the operational lifetime of the drive system. Identifying online the equivalent models of such drivelines can vastly improve the control effectiveness in terms of actively damping unwanted oscillations and can be used to monitor any abnormal behaviors that may lead to failures. More specifically, estimation of model parameters provides diagnostics and prognostics information, enabling predictive maintenance of the mechanical parts.

Control: There are several necessary research directions that are required to render power converters ready for future grid and process requirements, and a fresh look at the overall control structure that takes into account present and future requirements is an absolute must. Several key objectives like interoperability, grid support, overall process optimization, higher efficiency, and system availability need to be included into the design stage rather than adjusting existing control structures to handle requirements beyond what they were originally designed for. Even within a single power converter, coordinated control structures on the application layer are required, especially when multiple converters are connected through a process or to the same PCC.

Adaptive or reconfigurable control strategies have the potential to maximize the performance at each operating condition and grid or load status. This would require identifying the specific status of the grid, including fault mode, islanding or grid-connected mode, grid strength, or any sudden change in the generation or load. Supported with knowledge from the grid estimation, a controller should reconfigure to operate optimally under the various scenarios.

We have seen in Section 5.3 that the VSG is a good solution to handle future scenarios of grids being predominantly comprised of power electronic conversion systems. A more fundamental direction of research would require investigating better alternatives to VSG control that exploit the high dynamicity and versatility of power converters. Alternative approaches to be explored include constrained nonlinear control that takes into account natural constraints on currents and voltages and various system nonlinearities already in the design phase.

As the complexity of the system increases, data-driven control design [33, 34] may be the solution of choice for some control layers and this approach would augment the capability of model-based control design to tackle changing system behaviors in real-time.

Finally, it is important for any control structure or design approach to take into account the available computational capabilities onboard the embedded controllers being used in power conversion systems. This aspect is typically a significant limiting factor, especially since controls can utilize only a portion of the available memory and computational power of the embedded control hardware. Smart tailoring of any advanced control method to fit onboard the available control hardware is a key research direction that can enable the adoption of advanced methods versus traditional ones.

5.6 Conclusions

We are living in an exciting era and experiencing a paradigm shift in the way power conversion systems are required to operate, be it due to changes in the nature of the grid or in process requirements. We have reviewed in this chapter the impact of these changes both for grid-connected converters and for industrial drives. In both cases, we see a strong requirement for fundamental and applied research in power conversion control that also takes into account practical considerations such as system and operational limitations, as well as available computational capabilities for embedding advanced control methods. This work will pave the way to handle more complex type of grids or installations with multiple power conversion systems that interact with other types of storage and generation equipment. Moreover, it will allow the power conversion control community to move beyond the classical control approaches and explore the potential of advanced controls that are more versatile and adaptive in nature. This presents an incredible unique opportunity to shape a sustainable future for the society by increasing the efficiency and reliability in the energy and production sectors.

References

1 ENTSO-E Technical Group. High penetration of power electronic interfaced power sources and the potential contribution of grid forming converters. Technical report available online: https://euagenda.eu/upload/publications/untitled-292051-ea.pdf, 2020.

2 F. Milano, F. Doerfler, G. Hug, D.J. Hill, and G. Verbic. Foundations and challenges of low-inertia systems. *Power Systems Computation Conference (PSCC)*, pages 1–25, 2018. doi: 10.23919/PSCC.2018.8450880.

3 A. Monti, F. Milano, E. Bompard, and X. Guillaud. Converter-based dynamics and control of modern power systems.

4 P.T. Krein. *Elements of Power Electronics*. Oxford University Press, New York, 2015.

5 T. Geyer. *Model Predictive Control of High Power Converters and Industrial Drives*. Wiley, 2016.

6 S.-K. Sul. *Control of Electric Machine Drive Systems*. Wiley-IEEE Press, 2011.

7 D.G. Holmes and T.A. Lipo. *Pulse Width Modulation for Power Converters: Principles and Practice*. Wiley-IEEE Press, 2003.

8 P. Tiitinen and M. Surandra. The next generation motor control method, DTC direct torque control. In *Proceedings of International Conference on Power Electronics, Drives and Energy Systems for Industrial Growth*, New Delhi, India, volume 1, pages 37–43, 1996. doi: 10.1109/PEDES.1996.537279.

9 F. Blaabjerg. *Control of Power Electronic Converters and Systems*, volumes 1, 2, 3. Academic Press, 2018.

10 R. Teodorescu, M. Liserre, and P. Rodriguez. *Grid Converters for Photovoltaic and Wind Power Systems*. Wiley-IEEE Press, 2011.

11 IEEE. IEEE Standard for Interconnecting Distributed Resources with Electric Power Systems. *IEEE Std 1547-2003*, pages 1–28, 2003. doi: 10.1109/IEEESTD.2003.94285.

12 IEEE. IEEE Standard for Interconnection and Interoperability of Distributed Energy Resources with Associated Electric Power Systems Interfaces. *IEEE Std 1547-2018 (Revision of IEEE Std 1547-2003)*, pages 1–138, 2018. doi: 10.1109/IEEESTD.2018.8332112.

13 IEEE. IEEE Draft Recommended Practice for Establishing Methods and Procedures that Provide Supplemental Support for Implementation Strategies for Expanded Use of IEEE Standard 1547. *IEEE P1547.8/D8*, July 2014, pages 1–176, 2014.

14 ESO. GC0137: Minimum specification required for provision of GB Grid Forming (GBGF) capability (formerly Virtual Synchronous Machine (VSM) capability). *National Grid ESO*, 2022.

15 CLC/TC 9X. Railway Applications - Fixed Installations and Rolling Stock - Technical Criteria for the Coordination Between Electric Traction Power Supply Systems and Rolling Stock to Achieve Interoperability - Part 1: General. Standard, 2022.

16 IEEE Standard for Harmonic Control in Electric Power Systems. *IEEE Std 519-2022 (Revision of IEEE Std 519-2014)*, pages 1–31, 2022. doi: 10.1109/IEEESTD.2022.9848440.

17 S. Ansari, A. Chandel, and M. Tariq. A comprehensive review on power converters control and control strategies of AC/DC microgrid. *IEEE Access*, 9:17998–18015, 2021.

18 F. Doerfler, S. Bolognani, J.W. Simpson-Porco, and S. Grammatico. Distributed control and optimization for autonomous power grids. In *2019*

18th European Control Conference (ECC), pages 2436–2453, 2019. doi: 10.23919/ECC.2019.8795974.

19 D. Molzahn, F. Doerfler, H. Sandberg, S.H. Low, S. Chakrabarti, R. Baldick, and J. Lavaei. A survey of distributed optimization and control algorithms for electric power systems. *IEEE Transactions on Smart Grid*, 8(6):2941–2962, 2017.

20 Q. Peng, Q. Jiang, Y. Yang, T. Liu, H. Wang, and F. Blaabjerg. On the stability of power electronics-dominated systems: Challenges and potential solutions. *IEEE Transactions on Industry Applications*, 55(6):7657–7670, 2019. doi: 10.1109/TIA.2019.2936788.

21 M. Schweizer, S. Almer, S. Pettersson, A. Merkert, V. Bergemann, and L. Harnefors. Grid-forming vector current control. *IEEE Transactions on Power Electronics*, 37(11):13091–13106, 2022. doi: 10.1109/TPEL.2022.3177938.

22 H. Bevrani, T. Ise, and Y. Miura. Virtual synchronous generators: A survey and new perspectives. *International Journal of Electrical Power & Energy Systems*, 54:244–254, 2014.

23 S. d'Arco and J.A. Suul. Virtual synchronous machines - classification of implementations and analysis of equivalence to droop controllers for microgrids. In *2013 IEEE Grenoble Conference*, pages 1–7, Jun. 2013. doi: 10.1109/PTC.2013.6652456.

24 Adrian Gonzalez-Cajigas, Javier Roldan Perez, and Emilio Bueno. Design and analysis of parallel-connected grid-forming virtual synchronous machines for island and grid-connected applications. *IEEE Transactions on Power Electronics*, 37(5):5107–5121, 2021. doi: 10.1109/TPEL.2021.3127463.

25 API. Axial and Centrifugal Compressors and Expandercompressors. *API Standard 617*, 2022.

26 API. Tutorial on the API Standard Paragraphs Covering Rotor Dynamics and Balancing: An Introduction to Lateral Critical and Train Torsional Analysis and Rotor Balancing. *API Standard 684*, 2022.

27 AS. Demand Response Standard. *AS 4755 Standard*, 2020.

28 P. Hokayem and I. Pejcic. Model predictive damping of oscillations in an electrical converter system. US10404155B2, 2019.

29 P. Hokayem. Hybrid control method for an electrical converter. US20180062531A1, 2018.

30 A. Isaksson and S. Mastellone. Taming the power chain: How advanced control yields high availability and increased performance. *ABB Review*, 2015.

31 M. Mercangoez, S. Mastellone, S. Almer, T. Besselmann, P. Joerg, J. Niiranen, L. Peretti, and V.-M. Leppaenen. Damping torsional oscillations at the intersection of variable-speed drives and elastic mechanical systems. *ABB Review*, 2015.

32 P. Hokayem and P. Joerg. Damping torsional oscillations in a drive system. US11251742B2, 2022.

33 J. Coulson, J. Lygeros, and F. Dörfler. Data-enabled predictive control: In the shallows of the DeePC. In *European Control Conference*, pages 307–312, 2019. URL https://arxiv.org/abs/1811.05890.

34 L. Ortmann, J. Maeght, P. Panciatici, F. Dörfler, and S. Bolognani. Online feedback optimization for transmission grid operation, 2022. URL https://arxiv.org/abs/2212.07795. Submitted.

6

Robotics and Manufacturing Automation

Alisa Rupenyan[1] and Efe C. Balta[2]

[1]*ZHAW Centre for AI, School of Engineering, ZHAW Zurich University for Applied Sciences, Winterthur, Switzerland*
[2]*Control and Automation Group, inspire AG, Zurich, Switzerland*

6.1 Introduction

The digitalization of the industrial sector gives raise to complex demands on state-of-the-art manufacturing processes. In addition to the ever-present requirement for productivity, the manufacturing process of the future must be also energy efficient, sustainable, and offering the capability of lot size one production of individualized products. Achieving these goals requires highly automated, flexible, and reliably adapting production technology, while also optimizing for energy and material consumption. Crucial requirements for such systems are full plug-and-work functionality, fast reconfiguration, and fast adaption to different production goals [1]. Advanced automation solutions must enable self-learning capabilities in the control system, while maintaining stability and safety [2]. Existing manufacturing systems are already cyber-physical systems that generate a lot of data which is accessible to collect using IIoT sensing. Thus, manufacturing automation can be envisaged as spreading horizontally accross multiple industrial sectors: aeronautic, automotive, consumer goods, railway, and others. While industrial robotics solutions (i.e. robot arms) gain popularity, the other big innovation takes place within existing manufacturing systems, in applying advanced control and perception methods, borrowed from emerging robotics solutions.

The use of robotics in production is a key factor in making manufacturing economically viable. Robotics provides the means to reduce manufacturing unit costs. The market for manufacturing robots is strongly expected to grow through diversification into industries with lower volumes and into areas of

The Impact of Automatic Control Research on Industrial Innovation: Enabling a Sustainable Future,
First Edition. Edited by Silvia Mastellone and Alex van Delft.

manufacturing where manual assembly has previously moved away. Robots are also the key drivers of flexibility, adaptability, and reconfigurability. New automation concepts such as Human Robot Collaboration and Cyber-Physical Systems can impact and revolutionize the production landscape [3]. Increasing the flexibility of industrial robots and providing automation systems that offer faster, more intuitive configurations are essential goals for future production systems. The projected market for industrial robotics is expected to reach 75 billion USD by 2024, and the mark of 500,000 units installed per year worldwide is expected to be reached in 2024 [4].

6.1.1 Current Status/Challenges in Manufacturing

Research in robotics has seen incredible progress in the last years, driven by the incorporation of sensor data in the control algorithms of the robotic systems. The availability of data in this context has fueled the development of data-driven, learning-based control approaches. However, such exciting developments are somewhat slower adopted in modern manufacturing processes, mostly by incremental steps. Manufacturing technology innovation is largely dependent on the following factors in the deployed equipment and the specifics of its usage: (i) Large base and variability of existing equipment; (ii) Large variability in the deployed systems; (iii) Diverse manufacturing processes; (iv) Complex physics process interaction, which is not directly observable; (v) Real-time in situ sensing of process-induced structural changes remains a challenge.

The main requirements for the next-generation automated manufacturing systems are [1]:

- full plug-and-work functionality,
- fast reconfiguration,
- fast adaption to different production conditions and goals.

closely resembling those for autonomous systems in robotics. Learning has enabled self-driving [5], autonomous flight [6], and humanoid robots [7] to reach milestones that were previously unthinkable. For many advanced robotics applications, i.e. autonomous vehicles (AV), humanoid/walking robots, and unmanned aerial vehicles (UAV), the main challenges are to respond reliably to large uncertainties and unforeseen scenarios, which require self-optimization, evolving knowledge of the environment, and built-in safety mechanisms. Learning is already an indispensable component present in countless publications, and advances are driven by sensor fusion combined with machine learning and in particular reinforcement learning (RL), complementing advanced control algorithms. The technology roadmap for AVs [8], comprising six degrees of autonomy, serves as a mature example for automation in manufacturing. For AV,

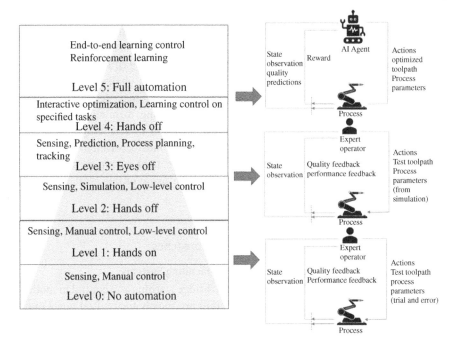

Figure 6.1 Levels of automation following the J3016 SAE standard on Automated driving levels, adapted for autonomous manufacturing in a single manufacturing system. Currently, the state of the manufacturing technology is predominantly around levels 1–2. Adopting innovative control technologies has the potential to bring it close to levels 4–5.

sensor technology is an integral part of autonomous robotic systems, comprising LIDARs, sensor fusion using IMUs, high dynamic range cameras, etc. The control solutions require an architecture unifying the information from sensor fusion, used for providing a visual map for orientation and planning, which serves as a basis to a path planning algorithm, connecting to the low-level controller to actuate the throttle and the steering wheel. Similarly, the six-level roadmap can be adapted to autonomous manufacturing systems, as attempted in Figure 6.1. In manufacturing, we receive feedback from the quality of the produced parts, which is often a separate process, involving separate equipment and human experts. We can think of planning the production path in terms of a tool path planning, controlling the process using feedback from integrated sensors and observers, and incorporating realistic process models for quality or equipment state predictions.

Although this parallel can be made, several differences arise in the manufacturing domain, mainly: (i) High variability/diversity in processes and associated equipment; (ii) Complex physics process interaction, which is not directly observable; (iii) Repeatability in the process (between batches, between parts, between layers); (iv) Lack of sensors at the production surface: Crucial to get real-time process information to control such systems for higher productivity and flexible manufacturing, but challenging to install sensors at the process spot. This challenge is addressed by recent tech enablers through edge/fog computing, however, real-time in situ sensing of process dynamics remains a challenge. (v) Lack of direct measurement of the microstructural evolution (anomaly detection): Specifically in material processing (e.g. additive manufacturing, forming, casting) even with real-time sensing, run-time data analytics, anomaly detection, and in situ part qualification is an important challenge which is mostly solved by destructive or postprocess testing. Even when there are sensors, they are mostly sensitive to immediate surface effects, and specific calibration using offline structural measurements is needed.

6.1.2 Existing Control Challenges and Approaches

6.1.2.1 Control Challenges in Existing Manufacturing Systems

Retrofitting While there are strong efforts to bridge learning and control in motion systems, the research efforts in this direction are mostly focused on developing or adapting RL for control [6]; providing safe boundaries and studying the role of uncertainty [9]; learning control for robotics [10]. What is less explored, is to develop and integrate learning methods for systems that are already in production and satisfy safety requirements, without introducing modifications to their hardware platforms. Such methods will enhance the productivity and enable existing systems to perform more complex tasks than those has been initially aimed for.

Lack of Real-Time Information on the Performance (Internal Structure Defects and Mechanical Properties Distribution) Due to the lack of adequate sensing and data acquisition on the plant floor, run-time performance assessment is often challenging. With robotics and manufacturing automation systems running continuously and undergoing wear and tear through their lifecycle, run-time analysis of the processes is key for understanding performance. There is a vast existing literature on fault diagnosis and identification (FDI) and prognostic health monitor [11, 12], and predictive maintenance (PdM) [13, 14]. However, most methods within this domain often rely on information that is either coming from idealized models, or statistical estimations that sacrifice performance in order to achieve classification robustness (e.g. minimizing false positives). Therefore, a current challenge is due to a lack of run-time representation of the specific physical resource of interest for further run-time analysis. To address this challenge, the use of digital twins has been proposed extensively in many areas of robotics and manufacturing automation [15, 16].

For many advanced processes, there is a lack of efficient in situ estimation and measurement methods due to the complexity of the physics involved. In-situ measurements are needed to evaluate the material properties, geometry, and mechanical performance of a product while it is being manufactured. As an example, in additive manufacturing (AM) processes, it is often difficult to perform characterization of the formed material in situ, which necessitates the use of ex-situ testing or even destructive testing. While there are a number of ways to address some of these issues in practice, the lack of such information in situ and in run-time makes it difficult develop efficient control methods to ensure desired material properties.

Accurate Models for Predictive Control Obtaining accurate depictions of the physical process is key for data analysis and control. Due to the interaction of multiple complex physics and nonlinearities, accurate modeling of robotics and manufacturing systems is challenging. Furthermore, as outlined earlier, models of the physical system may not reflect the run-time behavior and performance. Thus, accurate models for model-based control applications is an important existing challenge. Accurate process models often exist as computational models (e.g. [17–20]), which are often not suitable for control applications. Additionally, for certain processes such as additive manufacturing, dimensionality of spatiotemporal dynamical models often become a bottleneck for computation of efficient controllers [21]. In many cases, this dictates the use of model order reduction techniques [22], surrogate models of computationally intensive models [23, 24], or data-driven statistical models [25, 26]. While in many cases the accuracy of the resulting model is worse than the original approximation, there is an abundance of successful applications with feedback control in the literature. Another important challenge is thus uncertainty quantification for predictive models and answering the question: "Which model is good enough to deliver a desired performance objective via model-based control?"

Safety Another important consideration for control systems is the stringent safety specifications that need to be ensured by any controller running on a plant floor. Most industrial controllers employ a separate safety loop that can overrule all other actuation and decision-making mechanisms to immediately stop a process, e.g. dead-man switch on robotic teach pendants, light curtains on a production system to detect human intrusion, etc. While many static measures ensure operator safety of the system, there are many "soft" safety switches that are designed to ensure equipment health and safety. Since most of these safety sensors are designed around an operating point of the physical system, it is often challenging to implement controllers that perform set point changes, reconfigurations, or machine learning to improve process performance, as a safety controller may perceive such actions as nonsafe and overrule. Safe control synthesis and verification and validation methods aim to remedy this problem with several applications in the robotics and manufacturing automation domain [27]. However, without rigorous verification of safety, such controllers are difficult to implement in manufacturing systems with operators present. To implement intelligent controllers that can learn, adapt, and improve process performance, there is a clear need for efficient representation of safety and scalable methods for safety verification that are reliable.

Closed-Loop Process Optimization: Lack of Closed-Loop Process Control in Many Cases, Esp. Additive Manufacturing Advanced control techniques often rely on closed-loop optimization methods for improved process performance. In these control methods, the closed-loop performance of the process in run-time is formulated as an optimization problem to be solved by the controller to steer the process output in the desired manner. There are numerous examples of such closed-loop process optimization examples including those that incorporate learning-based controllers that either learn from the errors of past actions in repetitive settings [21, 28–32], or those that learn from data to improve the process model [33–38]. A common method of choice is model predictive control (MPC), where a model of the process is used for solving an optimization problem that minimizes a given cost function over a control and prediction horizon, such that the optimized inputs are expected to control the system for minimizing the cost function of interest. To improve robustness, the first optimal input is applied to the system, and the optimization is repeated in a receding horizon using new measurements. MPC-based approaches have been utilized in semiconductor manufacturing and in run-to-run control to utilize a process model to optimize control inputs [35, 39]. Additionally, process manufacturing and chemical processes have been the seminal applications of MPC in the past. However, widespread use of MPC for various processes and robotics-based automation requires fast and efficient computational resources, measurements, and communication infrastructures, which may be unavailable in current practice.

For many processes, run-time process measurements are either not available due to a lack of appropriate sensors, or in many cases, measurements collected from the system are not utilized for optimization-based closed-loop control purposes [40]. Optimization-based closed-loop control methods are often computationally demanding as an optimization problem (or an approximation of it) is solved at each time step. This means that depending on the application of interest, the control loop consisting of the measurement, computation, and actuation, needs to take place at high sampling rates, orders of kHz in some cases [41]. With the utilization of industrial IoT (IIoT) devices, acquisition of run-time data and computation are more accessible. However, for many modern manufacturing processes such as AM processes, run-time optimization-based control remains a challenge, not only due to the challenges in data collection, but also the previously mentioned challenges in this section [42, 43].

Advanced Operator Information Systems, Production, and Process Model-Based Systems to Support Operator Decisions A key element in the industrial automation systems is the human–machine interactions at various scales and interfaces. Many manufacturing processes rely on an operator to perform a desired operation or provide a reference, parameter set, etc., to specify the operation to be carried out by a machine. To specify the operation specifications, operators rely on physical feedback from the process (e.g. acoustics, visual inspection, etc.) and additional information support systems that provide data analytics. Information support systems analyze measurement data and historical data from the process to provide an overview of the run-time performance. In current practice, there is a lack of model-based information support systems that learn from the past behavior to improve their accuracy by adjusting to the changing process conditions. Instead, most information support systems simply analyze current measurements to provide an overview of the process history without too much insight on what it going on and let the operator draw conclusions from the presented data. The use of IIoT sensors, edge computing, artificial intelligence, and digital twins aim to address this issue by enabling a framework where the information support system continuously learns from the process to provide better predictions about the future behavior of the process, so that the operator is well informed when making decisions or adjusting process parameters.

System Architectures Software and hardware architectures in industrial processes are traditionally designed with the traditional automation and control solutions in mind. Within this setting, the control architecture is often rigid and has limited flexibility for modifications, both in terms of the controller tuning in run-time and in terms of run-time changes to the control architecture itself. Many industrial processes are run by traditional programmable logic controllers (PLCs) that may not provide flexibility to make such changes during process operation. Therefore, a practical challenge is to adapt industrial automation systems to architectures as reliable as traditional PLCs, but verifiable, programmable, and flexible with the state-of-the-art software packages for control at the same time. This fundamental challenge has attracted attention from fields outside of manufacturing systems as well. Solutions from networked device systems and interconnected autonomous agents are increasingly developed to be adapted as control and automation solutions for manufacturing systems. The industrial process controller of the future will incorporate advanced algorithms, artificial intelligence, and optimization in an efficient, reliable, and verifiable manner.

A common concern with increasing networking and autonomous controllers on the plant floor is cybersecurity of the physical resources. Cybersecurity threats emerge from a myriad of sources, ever increasing with more recent developments. Some of the most common sources of cyberattacks are network attacks that either inject fictitious communication to stun the communication of certain resources, or utilize vulnerable switches and structures in the network to cause harm. In the context of manufacturing, cyberattacks may have physical manifestation as well [44]. Additionally, the physical effect of a cyberattack may be difficult to detect in many scenarios [45, 46]. To address these issues, cyberattack detection and physics-based cyberattack detection has been a prominent recent research field. Many of these methods rely on sensory data from the physical system to model the normative system behavior and use run-time data to understand if the system is under attack [45, 47, 48]. Therefore, building controllers and autonomous systems that are secure and robust to such cyberattacks is an important challenge to address.

6.1.2.2 Additional Control Challenges for Specific Robotic Manufacturing Systems

Industrial robot arms, collaborative robots, and autonomous guided vehicles (AGVs) or other robots designed for manufacturing pose several specific challenges, which are provided in this section.

- Integration of cognitive functions into machines and robots for adaptability to changing manufacturing requirements: Industrial robotic systems are often designed to work under a given operating condition, for a certain task, etc. Therefore, moving toward industrial robots that learn about the changing tasks and context information around them to react accordingly is an important challenge to address. The complexity of analyzing cognitive changes and developing models to understand and react to those changes in a methodical manner is very high. Additionally, due to the uncertainty on the environment, especially for new tasks, adaptation to new tasks often comes with great conservatism that sacrifices performance. This challenge is related to the previous challenges on the lack of accurate models and learning system to understand the current operational state of the robot and its surroundings. Recent efforts focus on developing models that can be generalized and reused for similar tasks to mitigate safety and performance issues of industrial robots.
- Complex control systems involving the human in the control loop (HITL), supported by appropriate sensor systems and modeling and simulation tools; human robot interaction.

- Technical challenges: Accuracy of out of the box systems not sufficient for precise manufacturing (e.g. beyond 0.5 mm positioning error); not reprogrammable controllers which need to be circumvented by hierarchical control approaches; Existing command and communication delays in the system; insufficient state information requires installing new sensors and using sensor fusion methods.
- Data-handling: The intensive use of data requires data streaming solutions, and specific solutions for data sharing between manufacturers and equipment users, as well as established data standards, which are largely heterogeneous, or not well established.

The challenge of developing an accurate system model to apply model-based predictive control, iterative learning control (ILC), or other related methods, techniques can be overcome in learning control, where black-box models replace the mathematical models of the plant and the controllers, or first principles models are complemented by learning the deviations from the model [49–52]. Often, probabilistic modeling is preferred in the learning of the system dynamics, as it provides the associated uncertainty with the predictions, which can be included in the constraints of the underlying MPC-based controller [50]. A simple nominal model is often available in practice, while more complex nonlinearities can be challenging and time-intensive to model from first principles. MPC approach that integrates this nominal system with an additive part of the dynamics modeled as a Gaussian process (GP), where the resulting nonlinear stochastic control problem takes into account the model uncertainties associated with the GP has been demonstrated in [53]. This approach has been applied in the field of robotics (see, for example [54–56]), but to our knowledge, there are no demonstrations or adaptations to manufacturing and machining applications. Often resulting algorithms are showcased on available systems with flexible architecture, such as quadrotors [57], bipedal or quadrupedal robots [58], or advanced instrumentation for research purposes [53]. While these methods are promising, their real-time implementation requires either access to the controller of the system, or large interventions in the system architecture. Nonparametric learning models as GP regression require the availability of strong computational capabilities.

One successful strategy used in manufacturing and process control is to take advantage of the repetitive structure of the process and to correct for the performance error in a learning phase. ILC provides a framework that has been extensively explored for manufacturing problems [59–61]. Repetitive production operations provide the opportunity to use sensing and actuation monitoring to learn the effects of exogenous disturbances and complex dynamics on the process. Machine wear and feedstock variation are two typical manufacturing examples that require process control adaptation in order to maintain quality and throughput. ILC adapts the feedforward commands to the process machinery to compensate for the effects of such disturbances [62, 63]. In many cases, ILC requires access to the controller of the system. Broadening the scope of current ILC methods with learning [64–66] is key to maintaining the robustness of the manufacturing process, and to do it in a learning context requires a means of quantifying the level of variation present in the system. Probabilistic methods for learning the system performance, both as plant dynamics, and in closed loop, provide the necessary means to estimate and propagate the uncertainty in the process.

Combining the three classes of reference governor, predictive control, and ILC with learning facilitates achieving high productivity and to enables already existing machines to scale up. This approach is interesting both from practical point of view, providing enhanced performance of the machines, and from research point of view, exploring stability bounds and scalable implementation methods in learning-based control.

6.2 Vision

This section discusses the vision for the future of industrial automation from two main perspectives, machine-level, and system-level. The vision statements, technological enablers, and related trends are discussed in the following.

6.2.1 Machine Level

A further avenue for research emerging in manufacturing is purely learning-based approach to control, using end-to-end learning to control the system from raw sensor inputs to control actuation. This can be done using deep learning approaches, or RL. Furthermore, online learning techniques, where a near-optimal control action is learned from repetitions of the process or previous measurements, are utilized for improved system performance. Online learning techniques may be adopted in a purely data-driven fashion, where a system model is learned from the process data on-the-fly and a controller uses the model to optimize process performance. Alternatively, many existing methods utilize a nominal model of the process to then learn an optimal control action by repetitions of a specific task. The goal for many of these examples is to incorporate machine learning and data-driven approaches to go beyond robust control and further improve process performance. While data-driven and learning-based approaches alone are quite effective, incorporating subject matter expertise (SME) for synthesizing closed-loop controllers often yield improved performance in manufacturing processes [16, 67]. Adding SME knowledge may lead to understanding context changes, which is crucial for high-performance ML models [68]. Additionally, SME knowledge may be incorporated through defining process constraints and known physical relationships that may lead to improved phenomenological relations between process parameters, states, and outputs [47, 69–72].

For example, in autonomous racing, the performance of the car in the first lap is often unknown, and an MPC tracking-based online learning-control formulation can be applied to path planning only after the performance of the car following the first lap is known. Similar approaches for online learning control is often utilized in ILC, where an initial guess or even a zero input is applied to the system to initiate the learning loops. To draw parallels, in machining, we are interested in a high-precision tracking of a geometrical contour. In both cases, a system has to follow a given geometry, and to self-correct following an optimization-based procedure. There are examples of such learning control methods for precision motion stages, additive manufacturing, and many other industrial processes [21, 28–31]. The success of such applications motivates their utilization in next-generation industrial process control.

Another important direction for the future of machine-level technologies concerns monitoring applications for various purposes via online data analysis. Examples for such applications include predictive maintenance, anomaly detection, cybersecurity, and online verification. In today's industrial systems, monitoring, and online analysis modules are often decoupled from the control systems. Incorporating side information on the run-time operational state of the machine has the potential to improve controller performance by developing controllers that learn and adapt to the different operational states throughout the lifecycle of an industrial process. The paradigm of interconnecting the physical and cyber components of a system to develop a comprehensive framework for modeling, control, monitoring, and analysis is often framed as cyber-physical systems. Within the context of industrial processes, cyber-physical production systems, or cyber-physical manufacturing systems, are an emerging field in research aimed at developing necessary frameworks for the next generation industrial systems. Developing necessary monitoring technologies to complement automation systems is a necessary first step since control systems often rely on high-performance estimation and analysis modules to understand the behavior of the underlying physical system. Based on the measurements and online data analytics capabilities available, it is possible to generate models, at multiple fidelities, with the purpose of control, analysis, and verification.

Another important application of intelligent monitoring systems is to develop in situ verification for new reconfigurations and condition monitoring for process performance (e.g. anomaly detection, predictive maintenance, and cybersecurity). These aspects play a key role in deploying automation systems in practice since an autonomous system is expected to autonomously perform these analytics in a data-driven fashion. As an example, in a machining or additive manufacturing process, an in situ verification tool would use measurement data to assess if the resulting part conforms to the product specification, with a statistical uncertainty. As the controllers get more complex and learning-based methods are more integrated in to control architectures, such applications become crucial to ensure a satisfactory run-time performance. Similarly, understanding process performance and incorporating predictive maintenance to maximize process uptime and minimize unexpected downtime is a key goal for many resource critical applications (e.g. power plants and continuous processes). By introducing new capabilities for communication, sensing, and learning; cybersecurity becomes a major concern. Monitoring systems and integrated data analytics to detect and mitigate cyberattacks to manufacturing systems is an emergent field of research with numerous relevant applications in practice.

Putting the learning-based and data-driven controllers together with the smart monitoring systems results in the next generation of learning cyber-physical manufacturing systems that can flexibly adapt to changing conditions on the plant floor, and optimize their performance for a given task. We can summarise the framework for autonomous industrial manufacturing systems, drawing inspiration from control frameworks in robotics. In particular, in addition to the sensing and perception modules, indirect sensing (or so called soft sensing) needs to take place, by replacing the unobservable quantities using physics-based or ML models and observers, as shown on Figure 6.2. Learning an optimal setting and control inputs for optimal performance requires the controller to not only utilize the given information but also to explore the parameter space for other possible parameter combinations that are likely to yield better results. This fundamental problem is often named as the exploration-exploitation problem (E–E), where an agent either utilizes its knowledge about the environment to improve its performance in terms of a cost metric, or explores the environment to learn more and possibly find better solutions at the expense of worse cost. While the E–E problem for autonomous agents is a well studied problem in many fields, studying the same setting in manufacturing is challenging due to a number of reasons. First, due to the complex nature of the process, each manufacturing application poses unique challenges that are difficult to anticipate and solve ahead of time. Thus, an agent would need to explore a parameter set for each resource, which may be time consuming. Additionally, the exploration has to ensure product quality and constraints. Even when the product specifications are satisfied, it is often undesirable

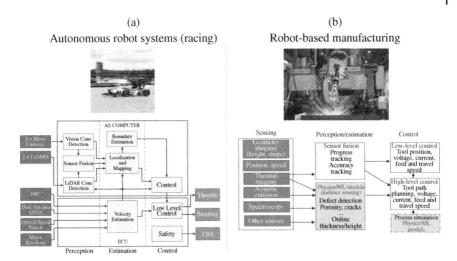

Figure 6.2 (a) Motion planning framework for autonomous systems. Source: Adapted from [73]. (b) Proposed framework for manufacturing. The new elements in the framework, specific to manufacturing (indirect sensing, process simulation/modeling), are highlighted with gray color.

to have fluctuations between products, which would be the case for an exploring controller on a process. Second, the parameter space for many manufacturing processes is often quite vast and many of the parameters are reconfigured based on changing context and product specifications. While operators are trained and often reliable to do such reconfiguration, an agent exploring the high-dimensional parameter space to find robust policies to do the same thing may be infeasible in practice. Therefore, an envisioned autonomous controller for the next generation cyber-physical manufacturing systems not only learns the best control action, but simultaneously explores better options, performs in situ verification, and balances the performance by considering maintenance cycles.

6.2.2 Production Level – Distributed Functionality

Existing production systems often utilize centralized decision-making to decide on a production plan to be executed in an open-loop fashion. The production plan is often executed by a manufacturing execution system (MES). A production plan that is optimized offline, or designed by a subject matter expert, is executed without run-time adjustments to account for disturbances. Although reconfiguration scenarios for disruptions such as unexpected downtime or a surge in demand levels are often utilized for extreme cases, open-loop execution of predetermined production plans often result in suboptimal performance.

An effective way to improve robustness to disturbances at a system-level is to utilize closed-loop controllers, e.g. scheduling and dispatch methods [74, 75], MPC [76–78], digital twin-based production scheduling [16, 79]. While many of these methods have shown to provide improved system performance, centralized methods often suffer from poor scalability in complex manufacturing systems. To remedy this, decentralized and distributed modeling, optimization, and control methods are utilized for closed-loop control of a manufacturing systems [80–83]. In the distributed setting, each resource in the system has some level of sensing and computation capacity at the edge level, often enabled through IIoT technology [84]. The distributed resources may have a modular structure that can be combined by a centralized resource [81], the edge computing may communicate with a centralized agent for optimization, or each distributed resource may have its own decision-maker, i.e. an autonomous agent [85].

The vision for the future of distributed control in manufacturing systems is to move toward autonomous agents that learn and cooperate on the plant floor to improve robustness and process performance, e.g. throughput, yield, quality, etc. [80, 81]. The challenges in computing, modeling, and cybersecurity become limiting factors in implementing such solutions as each agent is now an autonomous entity that communicates with the other agents in the system and requires computational resources to make decisions.

6.2.3 Formulation of Control Problems

Recently, several interesting directions in control approaches emerge in manufacturing, largely related to the common push for digitalization, amplified by initiatives such as Industry 4.0. While the use of cascade PID controllers is still predominant on a system-level [86], the adoption of more advanced control approaches such as MPC, ILC, and learning-based controllers increases, especially in high-precision systems [87], also fueled by increased computational capabilities of embedded controllers and edge devices, and by easier sensor integration.

Here we focus on approaches which unify relatively known methods and architectures with learning, and data augmentation. While multiple such approaches exist (e.g. data-enhanced predictive control, real-time (RT) feedback optimization, RL, etc.), we will look into the precision motion and positioning systems domain as example for industrial automation and robotics problems, in particular showcasing the use of digital twins for control design optimization. The advantage of digital twins is not only to simulate the manufacturing plant or system virtually but also to connect the physical interactions between process, system, and part quality using simulation models, process signals, and measurement results.

We study trajectory optimization and control of a precision motion system for the illustrative use case study in this section. Precision motion systems are widely used in many industrial processes including precision machining, semiconductor manufacturing, printing, and precision measurement [87]. The goal in precision motion control is to follow a given reference trajectory, e.g. the reference geometry in Figure 6.4, as well as possible in a given error metric.

Common error metrics include pointwise Euclidean distance to the reference path, or contour error that approximates the distance of two continuous curves [88]. The motion stage tasked with tracking a desired trajectory $\mathbf{r}_d$ with minimal contour error. The desired tracking geometry $\mathbf{r}_d$ is parametrized by the path parameter $s \in [0, L]$ with L as the total arc-length, so that we have $\mathbf{r}_d(s) = [r_{d,x}(s) \; r_{d,y}(s)]^\mathsf{T}$. We define $\bar{s} = s/L$ as the normalized path progression parameter. The true contour error $\hat{e}_k^c$ is the smallest perpendicular distance between the reference $\mathbf{r}_d(s)$ and output point p_k, i.e. the tool position. The corresponding path parameter $\hat{s}$ is given by $\hat{s} = \arg\min_s \|\mathbf{r}_d(s) - \mathbf{y}_k\|_2$. Then, the control objective is to design an input trajectory u such that $\hat{e}_k^c$ is minimized along the trajectory. These objectives are often posed as an optimization problem for minimizing the tracking error subject to the dynamics of the physical system and other input and output constraints for the process.

The dynamics of the process are represented by nonlinear discrete time dynamics

$$x_{k+1} = f(x_k, u_k) = f_0(x_k, u_k) + z_k, \tag{6.1}$$

$$p_k = g(x_k, u_k) = g_0(x_k, u_k) + w_k, \tag{6.2}$$

where $f : \mathbb{R}^{n_x} \times \mathbb{R}^{n_u} \to \mathbb{R}^{n_x}$ is the dynamics of the physical system, $g : \mathbb{R}^{n_x} \times \mathbb{R}^{n_u} \to \mathbb{R}^{n_p}$ is the measurement model of the system, and $f_0(\cdot, \cdot), g_0(\cdot, \cdot)$ are the nominal dynamics of the system given by

$$\tilde{x}_{k+1} = Ax_k + Bu_k + b_k := f_0(x_k, u_k), \tag{6.3}$$

$$\tilde{p}_k = Cx_k + Du_k := g_0(x_k, u_k), \tag{6.4}$$

for appropriately sized A, B, C, D, where $x_k = \tilde{x}_k + zk$ and $p_k = \tilde{p}_k + w_k$. We assume a linear measurement model for mapping the states onto the measured position

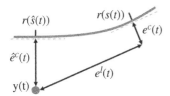

Figure 6.3 Illustration of the contour error on a reference geometry.

of the motion system at time k, e.g. for a 2D precision motion system we often have the position measurement of the tool tip in the Cartesian coordinates so that $p_k \in \mathbb{R}^2$.

The terms z_k and w_k represent the model mismatch terms for the state and output measurements. Depending on the assumption on these mismatch terms, we may formulate additional models for approximating their behavior to better model the dynamics of the system. Furthermore, these terms may include stochastic noise to model the process and measurement noise. As an example, if we assume that f, g are linear around a given operating point z_k, w_k are process and measurement noise. Then under suitable assumptions on the linear dynamics, we can use the well-known Kalman Filter (KF) techniques to estimate the state of the system from the measurements. When f, g are nonlinear but known, we can use nonlinear extensions such as extended and unscented KF methods to estimate states from the measurement.

In practice, the mismatch terms are assumed to be small, which is justified is the nominal control models f_0, g_0 are highly accurate. Under this assumption, the trajectory optimization problem is formulated as the following:

$$\min_{x,u} . \sum_{k=0}^{N-1} \left(x_k^\top Q x_k + u_k^\top R u_k + \Delta u_k^\top S \Delta u_k - f_u u_k \right) + x_N^\top Q_N x_N - f_{x,N} x_N, \quad (6.5a)$$

$$\text{s.t.:} \quad x_{k+1} = f_0(x_k, u_k) \tag{6.5b}$$

$$p_{i,k} = g_0(x_k, u_k) \tag{6.5c}$$

$$\Delta p_{i,k} \in [v_{\min} T, v_{\max} T], \qquad i = \{1, \ldots, n_a\} \tag{6.5d}$$

$$s_k \in [0, L], \ |e_k^l| \le e_{\max}^l, \ |e_k^c| \le e_{\max}^c. \tag{6.5e}$$

$$|e_N^l| \le \varphi^l e_{\max}^l, \quad |e_N^c| \le \varphi^c e_{\max}^c, \tag{6.5f}$$

$$x_0 = \hat{x}_0, \ u \in \mathcal{U} \tag{6.5g}$$

$$k = 0, \ldots, N-1,$$

where x, u are the state and input trajectories over the horizon N. The optimization problem has to trade off minimal tracking error with fast traversal of the geometry. Maneuver time and contour accuracy are essential in precision motion applications in order to achieve specified tolerances and maximize productivity. In the cost function (6.5a), they are balanced through the weights on the corresponding contour and lag errors in the state x, and on the progress weights linear in x and u [89, 90].

The dynamics model is provided in (6.5b). The remaining constraints are as follows: (6.5c)–(6.5d) denote a velocity constraints on the tool position as shown on Figure 6.3, (6.5e)–(6.5f) are constraints on the allowable contour error, longitudinal error, and path progression per horizon step, and (6.5g) defines the initial condition for the state vector, which is obtained by a state estimation scheme as discussed above, and additional input constraints are given through the convex set $\mathcal{U}^r$. The physical limitations of the machine impose velocity constraints in (6.5d) on the states, where n_a denotes the number of motion axes. $\varphi^{\{l,c\}} \in (0,1)$ enforces terminal error bounds. The input constraints include upper and lower bounds, and additional constraints on the first and second derivatives of the input change between consecutive horizon steps.

The problem in (6.5) describes a generic contour error optimization problem that can be solved offline to provide control inputs to the motion system in a feedforward fashion. Alternatively, the problem can be solved in a receding horizon fashion to provide appropriate control inputs at each time step, given appropriate choices of the problem parameters. While (6.5) above is well-studied in the past literature, there are several important challenges that remain. Since the motion system often operates at high frequencies at which it may not be feasible to solve the optimization problem (6.5) in a receding horizon. As a result, the control input to the motion system is often optimized in an iterative fashion where the controller optimizes an input trajectory, runs on the system, and re-optimizes the input based on the response of the system. However, in the offline trajectory optimization, the mismatch terms in the dynamics (6.1) may limit the achievable performance on the real system. Since the behavior of the physical system is often nonlinear, achieving desired performance in terms of tracking error and precision is often challenging.

An emerging method to overcome the issues on unpredictable online system behavior and difficult to model dynamics is to use learning control methods [34, 91]. Learning-based methods utilize input–output data of the system to fit statistical models that predict the true behavior of the system for control purposes, effectively modeling the mismatch terms in (6.1). While these methods are proven to be effective in practice, the necessary data collection and tuning may become a nontrivial task. Additionally, data-driven models may be nonrobust to context changes and reconfiguration, common in industrial settings. Digital twin technology has been emerging as a useful tool for addressing such challenges in optimization problems for control [16]. By learning the mismatch representations on the digital twin of the physical system instead of the real one, and tuning optimization and control parameters using the digital twin, it is possible to verify controllers prior to the deployment on the real system. Furthermore trajectories optimized using the digital twin are expected to provide similar performance on the real system, closing the gap between optimization and physical system.

Figure 6.4 illustrates the use of a Process Dynamics DT in several possible control loops. A digital twin of the physical motion stage is developed using input–output trajectories of the system, together with additional contextual information about the system. The Process Dynamics DT includes a model of

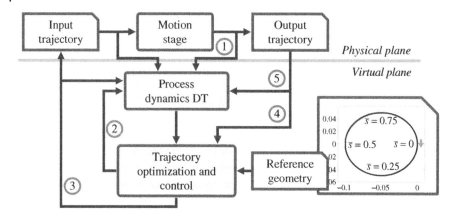

Figure 6.4 Digital Twin-based trajectory optimization and control loop. 1 – DT development, 2 – Trajectory optimization loop, 3 – Control loop, 4 – Measurement and data collection update, 5 – Continuous learning and performance assessment.

the system capable of simulating output response of the physical system, with additional computing resources, contextual data and database for model and additional process parameter storage and data interfaces to be used by other online analytics and control modules. The input–output model of the DT can be developed by a multitude of methods in the existing literature for high-fidelity nonlinear modeling. Additionally, data-driven components can be incorporated to improve the model accuracy. As DT is a live representation of the physical process, it needs to be capable of consuming process data to adapt and update its models over time. Notice that the modeling task here does not need to be control-oriented. In fact, complex and hierarchical models are often utilized as part of the DT to improve the prediction accuracy [16, 67].

After the development of the DT, the optimization problem (6.5) may be solved with the DT in the loop instead of the physical system. This way, the solutions may be computed either for the full trajectory, or in a receding horizon fashion since the DT can be run at any desired frequency to provide output predictions for a given input. The optimized trajectories can be fed to the real system to get optimal performance on the first run. The output trajectories of the physical system can be further used by an iterative controller to further tune the input trajectories using a framework similar to the one given in [91].

The model mismatch terms of (6.1) can be modeled in this context by exchanging the constraint (6.5b) with

$$x_{k+1} = f_0(x_k, u_k) + m(\theta|D), \tag{6.6}$$

where m is a function that approximates the mismatch of nominal model to the DT outputs using the data set D, parameterized by some set of features θ. A similar procedure can be incorporated for the output dynamics (6.5c). Note that now we exchange the task of learning the mismatch to the real system with learning the mismatch to the DT, which can be done efficiently and safely offline.

The Process Dynamics DT can be maintained and updated over time using the output measurements of the real system (step 5 in Figure 6.4). Additionally, the DT can be reconfigured and context changes can be efficiently encoded to update model parameters, simulation, and constraint parameters, as well as data formats. Therefore the Process Dynamics DT provides an up-to-date representation of the physical system for various data analytics and control purposes enabling the next generation of smart, context aware, and sustainable industrial processes.

Figure 6.5 shows the comparison of various input trajectory optimization methods on the given framework (Figure 6.4, number 3). $C1$ corresponds to using the reference geometry (the circle shown in Figure 6.5) as the input to the system, which corresponds to the case where no dynamic model of the physical system is used. $C2$ uses only the nominal model of the system to solve (6.5), with no usage of the DT of the system. The final case $C3$ makes use of the Process Dynamics DT to model the system as (6.6), using an efficient learning method for the mismatch term m in (6.6), named Bayesian Linear Regression (BLR). A similar approach for online mismatch learning with respect to a physical system is presented in [92]. We see that the mismatch model that uses the DT of the system outperforms all the other methods. The achieved improvement with respect to a controller without the mismatch term (using only the nominal dynamics model) is 30% in the RMS error and 18% in the maximum error.

Figure 6.6 presents the result of applying ILC in conjunction with mismatch learning methods using Gaussian Process Regression (GPR) as presented in [91]. By utilizing the nominal model and the mismatch model of the system, the controller iteratively optimizes the input trajectory to improve tracking performance.

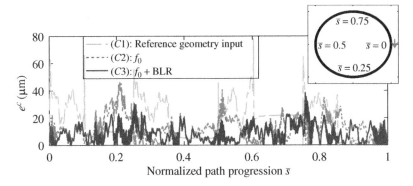

Figure 6.5 Experimental comparison of the contour error for different feedforward input trajectory optimization methods. BLR – Bayesian Linear Regression.

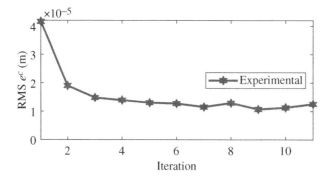

Figure 6.6 Experimental results for applying iterative learning control with mismatch learning methods. Source: [91]/IEEE.

This method can be used concurrently with the DT-based optimization methods, as shown in Figure 6.4 – 4. The combination of ILC with other DT-based learning optimization methods is expected to further improve the overall tracking performance of the physical system to enable control and learning loops at different levels.

6.3 Future Challenges and Trends

With the push to sustainability, energy and material resources use will become highly important, further enforcing optimization-based approaches in control systems used in industrial automation. Manufacturing, driven by individualization of production, and small lots demand, will require fast reconfigurability, associated with large computational capabilities. The underlying approaches will need to either be optimized for embedded controllers, or fast connection to cloud and edge systems, relying on fast communication channels, possibly enabled by 5G networks. This would also facilitate the coordination of autonomous production on multiple sites.

One of the most persistant trends in control for manufacturing systems, regardless if they are dedicated robotics systems, machining systems, or any hybrid between the two, is the emergence of digital twins for control, predictive and prescriptive maintenance, and process optimization. Importantly, they accommodate already deployed manufacturing systems with large variability, thus providing scalable solution for the current, as well as future equipment and operations. With the advent of fast communication networks, the role of DTs will be further expanded to assist synchronization between manufacturing, logistics, security, and quality assurance on a much larger scale than system level. The DTs integrate Internet of Things (IoT), big data, cloud and edge storage, artificial intelligence of things (AIoT), augmented reality (AR), etc. to form a comprehensive communication network for controlling, monitoring, diagnosis, and health inspection of equipment and facilities, traffic and transportation systems, buildings, and composite clusters. They have the potential to enable adaptive large-scale optimization-based approaches on factory and even large-scale infrastructural level, to achieve economic and sustainability objectives.

References

1 R.W. Brennan, P. Vrba, P. Tichy, A. Zoitl, C. Sander, T. Strasser, and V. Marik. Developments in dynamic and intelligent reconfiguration of industrial automation. *Computers in Industry*, 59(6):533–547, 2008. ISSN 0166-3615. doi: 10.1016/j.compind.2008.02.001. URL http://www.sciencedirect.com/science/article/pii/S0166361508000183.

2 Ray Y. Zhong, Xun Xu, Eberhard Klotz, and Stephen T. Newman. Intelligent manufacturing in the context of Industry 4.0: A review. *Engineering*, 3(5):616–630, 2017. ISSN 2095-8099. doi: 10.1016/J.ENG.2017.05.015. URL http://www.sciencedirect.com/science/article/pii/S2095809917307130.

3 Technological Roadmap. Smart advanced manufacturing. *Eureka Cluster technological roadmap*, 2019.

4 Industrial robotics market report, 2021. URL https://ifr.org/ifr-press-releases/news.

5 Eugenio Chisari, Alexander Liniger, Alisa Rupenyan, and John Lygeros. Learning from simulation, racing in reality. In *International Conference for Robotics and Automation (ICRA) 2021*, 2020. https://arxiv.org/abs/2011.13332.

6 Felix Berkenkamp, Matteo Turchetta, Angela P. Schoellig, and Andreas Krause. Safe model-based reinforcement learning with stability guarantees. In *Proceedings of the 31st International Conference on Neural Information Processing Systems*, NIPS'17, pages 908–919, USA, 2017. Curran Associates Inc. ISBN 978-1-5108-6096-4. URL http://dl.acm.org/citation.cfm?id=3294771.3294858.

7 S. Feng, E. Whitman, X. Xinjilefu, and C.G. Atkeson. Optimization based full body control for the Atlas robot. In *2014 IEEE-RAS International Conference on Humanoid Robots*, pages 120–127, 2014. doi: 10.1109/HUMANOIDS.2014.7041347.

8 Jesse Levinson, Jake Askeland, Jan Becker, Jennifer Dolson, David Held, Soeren Kammel, J. Zico Kolter, Dirk Langer, Oliver Pink, Vaughan Pratt, et al. Towards fully autonomous driving: Systems and algorithms. In *2011 IEEE Intelligent Vehicles Symposium (IV)*, pages 163–168. IEEE, 2011.

9 J.F. Fisac, A.K. Akametalu, M.N. Zeilinger, S. Kaynama, J. Gillula, and C.J. Tomlin. A general safety framework for learning-based control in uncertain robotic systems. *IEEE Transactions on Automatic Control*, 64(7):2737–2752, July 2019. ISSN 0018-9286. doi: 10.1109/TAC.2018.2876389.

10 Mohamed K. Helwa, Adam Heins, and Angela P. Schoellig. Provably robust learning-based approach for high-accuracy tracking control of Lagrangian systems. *IEEE Robotics and Automation Letters*, 4(2):1587–1594, 2019.

11 Ramin Moghaddass and Ming J. Zuo. An integrated framework for online diagnostic and prognostic health monitoring using a multistate deterioration process. *Reliability Engineering & System Safety*, 124:92–104, 2014.

12 W. Hu, A.G. Starr, and A.Y.T. Leung. Operational fault diagnosis of manufacturing systems. *Journal of Materials Processing Technology*, 133 (1-2):108–117, 2003.

13 Zeki Murat Çınar, Abubakar Abdussalam Nuhu, Qasim Zeeshan, Orhan Korhan, Mohammed Asmael, and Babak Safaei. Machine learning in predictive maintenance towards sustainable smart manufacturing in Industry 4.0. *Sustainability*, 12(19):8211, 2020.

14 Jinjiang Wang, Laibin Zhang, Lixiang Duan, and Robert X. Gao. A new paradigm of cloud-based predictive maintenance for intelligent manufacturing. *Journal of Intelligent Manufacturing*, 28(5):1125–1137, 2017.

15 Panagiotis Aivaliotis, Konstantinos Georgoulias, and George Chryssolouris. The use of digital twin for predictive maintenance in manufacturing. *International Journal of Computer Integrated Manufacturing*, 32(11):1067–1080, 2019.

16 James Moyne, Yassine Qamsane, Efe C. Balta, Ilya Kovalenko, John Faris, Kira Barton, and Dawn M. Tilbury. A requirements driven digital twin framework: Specification and opportunities. *IEEE Access*, 8:107781–107801, 2020.

17 K.-H. Leitz, P. Singer, A. Plankensteiner, B. Tabernig, H. Kestler, and L.S. Sigl. Multi-physical simulation of selective laser melting. *Metal Powder Report*, 72(5):331–338, 2017.

18 Saad A. Khairallah and Andy Anderson. Mesoscopic simulation model of selective laser melting of stainless steel powder. *Journal of Materials Processing Technology*, 214(11):2627–2636, 2014.

19 Efe C. Balta and Atakan Altınkaynak. Numerical and experimental analysis of bead cross-sectional geometry in fused filament fabrication. *Rapid Prototyping Journal*, 28(10):1882–1894, 2022.

20 Huanxiong Xia, Jiacai Lu, Sadegh Dabiri, and Gretar Tryggvason. Fully resolved numerical simulations of fused deposition modeling. Part I: Fluid flow. *Rapid Prototyping Journal*, 24(2):463–476, 2018. doi: 10.1108/RPJ-12-2016-0217.

21 David J. Hoelzle and Kira L. Barton. On spatial iterative learning control via 2-D convolution: Stability analysis and computational efficiency. *IEEE Transactions on Control Systems Technology*, 24(4):1504–1512, 2015.

22 Wilhelmus H.A. Schilders, Henk A. Van der Vorst, and Joost Rommes. *Model order reduction: Theory, research aspects and applications*, volume 13. Springer, 2008.

23 Kevin McBride and Kai Sundmacher. Overview of surrogate modeling in chemical process engineering. *Chemie Ingenieur Technik*, 91(3):228–239, 2019.

24 Dirk Gorissen, Ivo Couckuyt, Piet Demeester, Tom Dhaene, and Karel Crombecq. A surrogate modeling and adaptive sampling toolbox for computer based design. *Journal of Machine Learning Research.-Cambridge, Mass.*, 11:2051–2055, 2010.

25 Samuel Balula, Dominic Liao-McPherson, Alisa Rupenyan, and John Lygeros. Data-driven reference trajectory optimization for precision motion systems. *arXiv preprint arXiv:2205.15694*, 2022.

26 Bowen Wang, Guobin Zhang, Huizhi Wang, Jin Xuan, and Kui Jiao. Multi-physics-resolved digital twin of proton exchange membrane fuel cells with a data-driven surrogate model. *Energy and AI*, 1:100004, 2020.

27 Andreas Löcklin, Manuel Müller, Tobias Jung, Nasser Jazdi, Dustin White, and Michael Weyrich. Digital twin for verification and validation of industrial automation systems–a survey. In *2020 25th IEEE International Conference on Emerging Technologies and Factory Automation (ETFA)*, volume 1, pages 851–858. IEEE, 2020.

28 Hyo-Sung Ahn, YangQuan Chen, and Kevin L. Moore. Iterative learning control: Brief survey and categorization. *IEEE Transactions on Systems, Man, and Cybernetics, Part C (Applications and Reviews)*, 37(6):1099–1121, 2007.

29 David H. Owens and Jari Hätönen. Iterative learning control—an optimization paradigm. *Annual Reviews in Control*, 29(1):57–70, 2005.

30 Dominic Liao-McPherson, Efe C. Balta, Alisa Rupenyan, and John Lygeros. On robustness in optimization-based constrained iterative learning control. *IEEE Control Systems Letters*, 6:2846–2851, 2022.

31 Jurgen Van Zundert, Joost Bolder, and Tom Oomen. Optimality and flexibility in iterative learning control for varying tasks. *Automatica*, 67:295–302, 2016.

32 Ugo Rosolia and Francesco Borrelli. Learning model predictive control for iterative tasks. A data-driven control framework. *IEEE Transactions on Automatic Control*, 63(7):1883–1896, 2017.

33 Tor Aksel N. Heirung, B. Erik Ydstie, and Bjarne Foss. Dual adaptive model predictive control. *Automatica*, 80:340–348, 2017.

34 Lukas Hewing, Juraj Kabzan, and Melanie N. Zeilinger. Cautious model predictive control using Gaussian process regression. *IEEE Transactions on Control Systems Technology*, 28(6):2736–2743, 2019.

35 Aftab A. Khan, James R. Moyne, and Dawn M. Tilbury. Virtual metrology and feedback control for semiconductor manufacturing processes using recursive partial least squares. *Journal of Process Control*, 18(10):961–974, 2008.

36 Salvatore Nicosia and Patrizio Tomei. Model reference adaptive control algorithms for industrial robots. *Automatica*, 20(5):635–644, 1984.

37 John J. Craig, Ping Hsu, and S. Shankar Sastry. Adaptive control of mechanical manipulators. *The International Journal of Robotics Research*, 6(2):16–28, 1987.

38 Hanlei Wang. Adaptive control of robot manipulators with uncertain kinematics and dynamics. *IEEE Transactions on Automatic Control*, 62 (2):948–954, 2016.

39 James Moyne, Enrique Del Castillo, and Arnon M. Hurwitz. *Run-to-run control in semiconductor manufacturing*. CRC Press, 2018.

40 Kira Barton, Doug Bristow, David Hoelzle, and Sandipan Mishra. Mechatronics advances for the next generation of AM process control. *Mechatronics*, 64(C):102281, 2019.

41 Volker Renken, Axel von Freyberg, Kevin Schünemann, Felix Pastors, and Andreas Fischer. In-process closed-loop control for stabilising the melt pool temperature in selective laser melting. *Progress in Additive Manufacturing*, 4(4):411–421, 2019.

42 Dominic Liao-McPherson, Efe C. Balta, Ryan Wüest, Alisa Rupenyan, and John Lygeros. In-layer thermal control of a multi-layer selective laser melting process. In *2022 European Control Conference (ECC)*, pages 1678–1683. IEEE, 2022.

43 Riccardo Zuliani, Efe C. Balta, Alisa Rupenyan, and John Lygeros. Batch model predictive control for selective laser melting. In *2022 European Control Conference (ECC)*, pages 1560–1565. IEEE, 2022.

44 Sofia Belikovetsky, Mark Yampolskiy, Jinghui Toh, Jacob Gatlin, and Yuval Elovici. dr0wned– {Cyber-Physical} attack with additive manufacturing. In *11th USENIX Workshop on Offensive Technologies (WOOT17)*, 2017.

45 Priyanka Mahesh, Akash Tiwari, Chenglu Jin, Panganamala R. Kumar, A.L. Narasimha Reddy, Satish T.S. Bukkapatanam, Nikhil Gupta, and Ramesh Karri. A survey of cybersecurity of digital manufacturing. *Proceedings of the IEEE*, 109(4):495–516, 2020.

46 Fabio Pasqualetti, Florian Dörfler, and Francesco Bullo. Attack detection and identification in cyber-physical systems. *IEEE Transactions on Automatic Control*, 58(11):2715–2729, 2013.

47 Jairo Giraldo, David Urbina, Alvaro Cardenas, Junia Valente, Mustafa Faisal, Justin Ruths, Nils Ole Tippenhauer, Henrik Sandberg, and Richard Candell.

A survey of physics-based attack detection in cyber-physical systems. *ACM Computing Surveys (CSUR)*, 51(4):1–36, 2018.

48 Efe C. Balta, Michael Pease, James Moyne, Kira Barton, and Dawn M. Tilbury. Digital twin-based cyber-attack detection framework for cyber-physical manufacturing systems. *IEEE Transactions on Automation Science and Engineering*, 2023. doi: 10.1109/TASE.2023.3243147.

49 B. Bócsi, P. Hennig, L. Csató, and J. Peters. Learning tracking control with forward models. In *2012 IEEE International Conference on Robotics and Automation*, pages 259–264, May 2012. doi: 10.1109/ICRA.2012.6224831.

50 J. Hall, C. Rasmussen, and J. Maciejowski. Modelling and control of nonlinear systems using Gaussian processes with partial model information. In *2012 IEEE 51st IEEE Conference on Decision and Control (CDC)*, pages 5266– 5271, Dec. 2012. doi: 10.1109/CDC.2012.6426746.

51 Chris J. Ostafew, Angela P. Schoellig, and Timothy D. Barfoot. Robust constrained learning-based NMPC enabling reliable mobile robot path tracking. *The International Journal of Robotics Research*, 35(13):1547–1563, 2016. doi: 10.1177/0278364916645661.

52 M.P. Deisenroth, D. Fox, and C.E. Rasmussen. Gaussian processes for data-efficient learning in robotics and control. *IEEE Transactions on Pattern Analysis and Machine Intelligence*, 37(2):408–423, Feb. 2015. ISSN 0162-8828. doi: 10.1109/TPAMI.2013.218.

53 E.D. Klenske, M.N. Zeilinger, B. Schölkopf, and P. Hennig. Gaussian process-based predictive control for periodic error correction. *IEEE Transactions on Control Systems Technology*, 24(1):110–121, Jan. 2016. ISSN 1063-6536. doi: 10.1109/TCST.2015.2420629.

54 G. Cao, E.M. Lai, and F. Alam. Gaussian process model predictive control of unknown non-linear systems. *IET Control Theory Applications*, 11(5):703–713, 2017. ISSN 1751-8644. doi: 10.1049/iet-cta.2016.1061.

55 David B. Grimes, Rawichote Chalodhorn, and Rajesh P.N. Rao. Dynamic imitation in a humanoid robot through nonparametric probabilistic inference. In *Proceedings of Robotics: Science and Systems (RSS'06)*. MIT Press, 2006.

56 N. Banka and S. Devasia. Application of iterative machine learning for output tracking with magnetic soft actuators. *IEEE/ASME Transactions on Mechatronics*, 23(5):2186–2195, Oct. 2018. ISSN 1083-4435. doi: 10.1109/TMECH.2018.2855217.

57 Felix Berkenkamp, Angela P. Schoellig, and Andreas Krause. Safe controller optimization for quadrotors with Gaussian processes. In *Proceedings of the International Conference on Robotics and Automation (ICRA)*, pages 491–496, May 2016.

58 Dingsheng Luo, Yaoxiang Ding, Xiaoqiang Han, Yang Ma, Yian Deng, Zhan Liu, and Xihong Wu. Humanoid environmental perception with Gaussian process regression. *International Journal of Advanced Robotic Systems*, 13(6):1729881416666783, 2016. doi: 10.1177/1729881416666783.

59 D.A. Bristow, M. Tharayil, and A.G. Alleyne. A survey of iterative learning control. *IEEE Control Systems Magazine*, 26(3):96–114, June 2006. ISSN 1066-033X. doi: 10.1109/MCS.2006.1636313.

60 Titus Haas, Natanael Lanz, Roman Keller, Sascha Weikert, and Konrad Wegener. Iterative learning for machine tools using a convex optimisation approach. *Procedia CIRP*, 46:391–395, 2016. ISSN 2212-8271. doi: 10.1016/j.procir.2016.04.033. URL http://www.sciencedirect.com/science/article/pii/S2212827116301883. 7th HPC 2016 CIRP Conference on High Performance Cutting.

61 M. Tsai, C. Yen, and H. Yau. Integration of an empirical mode decomposition algorithm with iterative learning control for high-precision machining. *IEEE/ASME Transactions on Mechatronics*, 18(3):878–886, June 2013. ISSN 1083-4435. doi: 10.1109/TMECH.2012.2194162.

62 F.L. Mueller, A.P. Schoellig, and R. D'Andrea. Iterative learning of feed-forward corrections for high-performance tracking. In *2012 IEEE/RSJ International Conference on Intelligent Robots and Systems*, pages 3276–3281, Oct. 2012. doi: 10.1109/IROS.2012.6385647.

63 T. Ravensbergen, P.C. de Vries, F. Felici, T.C. Blanken, R. Nouailletas, and L. Zabeo. Density control in ITER: An iterative learning control and robust control approach. *Nuclear Fusion*, 58(1):016048, Dec. 2017. doi: 10.1088/1741-4326/aa95ce.

64 P. Jiang, L.C.A. Bamforth, Z. Feng, J.E.F. Baruch, and Y. Chen. Indirect iterative learning control for a discrete visual servo without a camera-robot model. *IEEE Transactions on Systems, Man, and Cybernetics, Part B (Cybernetics)*, 37(4):863–876, Aug. 2007. ISSN 1083-4419. doi: 10.1109/TSMCB.2007.895355.

65 B. Nemec, M. Simonič, N. Likar, and A. Ude. Enhancing the performance of adaptive iterative learning control with reinforcement learning. In *2017 IEEE/RSJ International Conference on Intelligent Robots and Systems (IROS)*, pages 2192–2199, Sep. 2017.

66 Samuel Balula, Efe C. Balta, Dominic Liao-McPherson, Alisa Rupenyan, and John Lygeros. Sequential quadratic programming-based iterative learning control for nonlinear systems. In *2023 IEEE Conference on Control Technology and Applications (CCTA)*, pages 162–167. IEEE, 2023. doi: 10.1109/CCTA54093.2023.10253186.

67 Yassine Qamsane, James Moyne, Maxwell Toothman, Ilya Kovalenko, Efe C. Balta, John Faris, Dawn M. Tilbury, and Kira Barton. A methodology to develop and implement digital twin solutions for manufacturing systems. *IEEE Access*, 9:44247–44265, 2021.

68 Juan Pablo Usuga Cadavid, Samir Lamouri, Bernard Grabot, Robert Pellerin, and Arnaud Fortin. Machine learning applied in production planning and control: A state-of-the-art in the era of Industry 4.0. *Journal of Intelligent Manufacturing*, 31(6):1531–1558, 2020.

69 Rolf Isermann. Model-based fault-detection and diagnosis– status and applications. *Annual Reviews in control*, 29(1):71–85, 2005.

70 Efe C. Balta, Dawn M. Tilbury, and Kira Barton. A digital twin framework for performance monitoring and anomaly detection in fused deposition modeling. In *2019 IEEE 15th International Conference on Automation Science and Engineering (CASE)*, pages 823–829. IEEE, 2019.

71 Dan Li, Nagi Gebraeel, and Kamran Paynabar. Detection and differentiation of replay attack and equipment faults in SCADA systems. *IEEE Transactions on Automation Science and Engineering*, 18(4):1626–1639, 2020.

72 Uduak Inyang-Udoh and Sandipan Mishra. A physics-guided neural network dynamical model for droplet-based additive manufacturing. *IEEE Transactions on Control Systems Technology*, 30(5):1863–1875, 2021.

73 José L. Vázquez, Marius Brühlmeier, Alexander Liniger, Alisa Rupenyan, and John Lygeros. Optimization-based hierarchical motion planning for autonomous racing. *accepted in IEEE IROS 2020*, 2020. https://arxiv.org/abs/2003.04882.

74 Jian Zhang, Guofu Ding, Yisheng Zou, Shengfeng Qin, and Jianlin Fu. Review of job shop scheduling research and its new perspectives under Industry 4.0. *Journal of Intelligent Manufacturing*, 30(4):1809–1830, 2019.

75 Yassine Qamsane, Efe C. Balta, James Moyne, Dawn Tilbury, and Kira Barton. Dynamic rerouting of cyber-physical production systems in response to disruptions based on SDC framework. In *2019 American Control Conference (ACC)*, pages 3650–3657. IEEE, 2019.

76 Efe C. Balta, Mohammad H. Mamduhi, John Lygeros, and Alisa Rupenyan. Controller-aware dynamic network management for Industry 4.0. *arXiv preprint arXiv:2205.14449*, 2022.

77 Efe C. Balta, Ilya Kovalenko, Isaac A. Spiegel, Dawn M. Tilbury, and Kira Barton. Model predictive control of priced timed automata encoded with first-order logic. *IEEE Transactions on Control Systems Technology*, 30(1):352–359, 2021.

78 Michael Baldea, Juan Du, Jungup Park, and Iiro Harjunkoski. Integrated production scheduling and model predictive control of continuous processes. *AIChE Journal*, 61(12):4179–4190, 2015.

79 Yassine Qamsane, Chien-Ying Chen, Efe C. Balta, Bin-Chou Kao, Sibin Mohan, James Moyne, Dawn Tilbury, and Kira Barton. A unified digital twin framework for real-time monitoring and evaluation of smart manufacturing systems. In *2019 IEEE 15th International Conference on Automation Science and Engineering (CASE)*, pages 1394–1401. IEEE, 2019.

80 Ilya Kovalenko, Efe C. Balta, Dawn M. Tilbury, and Kira Barton. Cooperative product agents to improve manufacturing system flexibility: A model-based decision framework. *IEEE Transactions on Automation Science and Engineering*, 20(1):440–457, 2022.

81 Ilya Kovalenko, James Moyne, Mingjie Bi, Efe C. Balta, Wenyuan Ma, Yassine Qamsane, Xiao Zhu, Z. Morley Mao, Dawn M. Tilbury, and Kira Barton. Toward an automated learning control architecture for cyber-physical manufacturing systems. *IEEE Access*, 10:38755–38773, 2022.

82 Ilya Kovalenko, Dawn Tilbury, and Kira Barton. The model-based product agent: A control oriented architecture for intelligent products in multi-agent manufacturing systems. *Control Engineering Practice*, 86:105–117, 2019.

83 Birgit Vogel-Heuser, Jay Lee, and Paulo Leitão. Agents enabling cyber-physical production systems. *at-Automatisierungstechnik*, 63(10):777–789, 2015. doi: 10.1515/auto-2014-1153.

84 Goiuri Peralta, Markel Iglesias-Urkia, Marc Barcelo, Raul Gomez, Adrian Moran, and Josu Bilbao. Fog computing based efficient IoT scheme for the Industry 4.0. In *2017 IEEE International Workshop of Electronics, Control, Measurement, Signals and their Application to Mechatronics (ECMSM)*, pages 1–6, 2017. doi: 10.1109/ECMSM.2017.7945879.

85 Paulo Leitão. Agent-based distributed manufacturing control: A state-of-the-art survey. *Engineering Applications of Artificial Intelligence*, 22(7):979–991, 2009. ISSN 0952-1976. doi: 10.1016/j.engappai.2008.09.005. URL https://www.sciencedirect.com/science/article/pii/S0952197608001437. Distributed Control of Production Systems.

86 Anca Maxim, Dana Copot, Cosmin Copot, and Clara M. Ionescu. The 5w's for control as part of industry 4.0: Why, what, where, who, and when—a PID and MPC control perspective. *Inventions*, 4(1):10, 2019.

87 Tom Oomen. Advanced motion control for precision mechatronics: Control, identification, and learning of complex systems. *IEEJ Journal of Industry Applications*, 7(2):127–140, 2018.

88 Yoram Koren and Ch-Ch Lo. Variable-gain cross-coupling controller for contouring. *CIRP Annals*, 40(1):371–374, 1991.

89 Samuel Balula, Alex Liniger, Alisa Rupenyan, and John Lygeros. Reference design for closed loop system optimization. In *IEEE Xplore European Control Conference*, 2020.

90 Alisa Rupenyan, Mohammad Khosravi, and John Lygeros. Performance-based trajectory optimization for path following control using Bayesian optimization. In *2021 60th IEEE Conference on Decision and Control (CDC)*, pages 2116–2121, 2021. doi: 10.1109/CDC45484.2021.9683482.

91 Efe C. Balta, Kira Barton, Dawn M. Tilbury, Alisa Rupenyan, and John Lygeros. Learning-based repetitive precision motion control with mismatch compensation. In *IEEE Conference on Decision and Control*, 2021.

92 Christopher D. McKinnon and Angela P. Schoellig. Learn fast, forget slow: Safe predictive learning control for systems with unknown and changing dynamics performing repetitive tasks. *IEEE Robotics and Automation Letters*, 4(2):2180–2187, 2019.

7

Process Industry
Alex van Delft

VanDelft.IT, Sittard, The Netherlands

7.1 Introduction

In this chapter, the framework of Chapter 1 is applied to the Process Industry sector. First, this sector is defined and introduced, and background information is provided. The current stand of technology and innovation related to the sector is described. We will use two illustrative examples throughout this chapter to highlight the concepts presented: a case originating from a continuous process plant and a batch industry case. In Section 7.2, existing control challenges as perceived in the industry are listed. Section 7.3 discusses the current challenges and vision of the future generation of products and processes. In Section 7.4, control problems and research directions for process industry will be formulated. Section 7.5 will conclude this chapter, focusing the question: what drives and blocks innovation in the process industry, and what could be done about this?

7.1.1 Definition of Process Industry

Process Industry, in contrast to discrete industries, comprises the processing of raw materials into finished products. In the framework of this volume, we consider the following specific process industry categories: Oil and Gas, Metals and Mining, Chemicals, Pharmaceuticals, Biotech processes, Pulp and Paper, Food and Beverages, Water, and Wastewater. This broad sector presents distinct differences in asset intensity, production flexibility, etc., and as such, also in control challenges and potential for innovation. In many cases, however, it is possible to define common directions in control research and innovation. In this volume, we will focus mainly on the chemicals industry, and use this subsector to describe control challenges and research directions. It should be noted, however, that the approach and

The Impact of Automatic Control Research on Industrial Innovation: Enabling a Sustainable Future, First Edition. Edited by Silvia Mastellone and Alex van Delft.

also many of the observations are to a large extent applicable to other domains in process industry.

Processing raw materials into finished products has an impressive history and can be regarded as almost as old as mankind. Examples are cooking food, production of bricks, and melting metals. The big developments in this industry, including the need for associated controls, occurred after the invention of the steam engine and the following industrial revolution. A further acceleration originated from the use of fossil fuels at the end of the nineteenth century, enabling the oil and gas and petrochemical industries.

Process industries are characterized by complex and rigid supply and value chains [1], and products consisting of materials or ingredients rather than components [2].

The typical production processes consist of distinct process steps (unit operations), e.g. reaction, separation/purification, extrusion, homogenization, drying, etc. Traditionally, these steps each have their own input/output requirements regarding quality and reproducibility. Orchestration of these unit operations would yield a better overall result but it is still a challenge in many of the process industries. For the sake of simplicity, we exclude production steps like tableting, packaging from this analysis, as these are discrete production steps.

The impact of this industry sector can be described in terms of worldwide turnover, people employed, energy consumption, and emissions. For a highly industrialized country like The Netherlands, key figures are depicted in Figure 7.1. Worldwide capital spending for the chemicals process industry has more than tripled from 2005 to 2015, despite the financial crisis, and is showing a remarkable change in geographical focus over the last 10 years, according to the European Chemical Industry Council (CEFIC) [3].

With its impact on energy consumption and emissions, the process industry will focus in the next 10–20 years mainly on moving from fossil fuels to electricity (energy transition) and further reduction of the ecofootprint, according to the UN's sustainable development goals. This is expected to have major consequences for

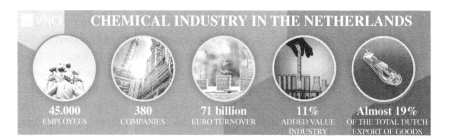

Figure 7.1 Key figures of the process industry in The Netherlands. Source: www.vnci.nl.

the process technologies applied. And will be the main driver for future innovation in this sector.

A conceptual view of a process industry normally comprises three axes [4]: (i) The Value chain: the physical process from raw materials to product. (ii) The Asset life cycle: from plant design to operations and maintenance. (iii) The Vertical integration process: from plant floor to production and business management. Each of these axes has its own challenges with respect to automatic control.

Currently, in Process Industries, the Innovation process is highly influenced by:

- Process and product safety: In most process industries, the assessment of the safety impact on the production process as well as end product is the first and foremost when judging an innovative process (control) technology: the "License to Operate" needs to be preserved. In practice, this often leads to an "if in doubt, don't" result: if the impact is unknown or difficult to measure, the innovation is blocked. This is enhanced by the fact that the possibilities to do trial runs or experiments in process installations are often very limited. On the other hand, safety without reliable operating basic control is unthinkable, as history has shown. Innovative opportunities exist in more advanced sensing, online fault detection, data analytics/AI.
- Quality and delivery reliability: Producing the right product quality in the right quantity at the right time is obviously one of the most important Operations Metrics in process industry, where automatic control can play a key role. Handling the impact of variations in raw materials and process variability is a domain where automatic control is a well-established partner, albeit most often on a basic level. A limiting factor is often the lack of accurate process knowledge represented in models. Process Analytical Technology, a term often used in Pharma, combines sensor knowledge with model relationships. In many process industries, a separate quality department is responsible for end-product quality, which in fact results in an extra potential constraint when introducing process innovations. Concerning delivery reliability: an important aspect is the avoidance of intermediate and end products stocks since this is mostly wasted capital. To achieve this, the aspect of "orchestration" of process units comes into play. And for a salesperson it is important to be able to provide a sharp "available to promise": being able to sell a product to be delivered at the right time and place, in the right quantity and quality.
- Compliance with governmental regulations: Most governments impose requirements on process industry regarding the impact on safety of people (as mentioned above) and the impact on the environment. The latter is measured in terms of targets in reduction of energy consumption, reduction of (carbon or VOC) emission, water use, impact on open water, etc. Innovations in this area will become key in the next decennia. For especially the pharma/food industries,

governmental regulations like FDA, the US Food and Drug Administration [5], are impacting and sometimes limiting the innovation process: any change in the production process, including the introduction of, e.g. advanced controls, requires a re-qualification or validation, which is a time-consuming and cumbersome process. On the other hand, the same regulations require advanced tracking and tracing (showing that the production process was "in control").

- Economic considerations: In addition to the quality aspect mentioned above, which also has an economic impact, this is the traditional main yardstick for control innovations: "show me the business case." Most often followed with the remark: "make sure it is proven in use." For that matter, it is thanks to the innovative spirit of some petrochemical companies that Model Predictive Control became a defacto standard in that industry.
- Innovations are being hampered by current assets: This can be regarded as a general law in many parts of process industry. Investments in production assets are quite often huge and result in plant life cycles of sometimes 30-40 years, during which innovations are limited. In a plant turnaround (typically every 4–5 years in petro- and bulkchemicals), process innovations can be implemented including the associated controls. The reality, however, is that because of the strict management of turnaround times including the required safety studies, there is hardly an opportunity for experimentation and implementation of advanced controls. And due to the focus on asset reliability, the replacement of an obsolete process control system (which would be an excellent opportunity to introduce more advanced controls) most often results in a "like for like" replacement.

Innovation in process industries may take place across the complete value chain. Automatic control innovations need to find their position in the complete innovation process, as highlighted by Lager [2, 6].

7.1.2 Automatic Control in Process Industry: Current Stand

People-Processes-Tools/Systems

If we look at the contribution of automatic control in an industry sector, we cannot limit it to the technology aspect only. As argued in [7], successful and sustainable implementation of innovations in a production organization requires People/Processes/Tools/Systems, as depicted in Figure 7.2. Technology is of course the major player that delivers value. However, to enable this innovation process the people working with the technology need to be properly instructed, and the work processes regarding the technology must be properly arranged. To give one example: Introducing Model Predictive Control [8, 9] in a chemical plant is bound for failure, if the operators have not been engaged in the implementation process: they may switch off the controller anytime if they don't understand or

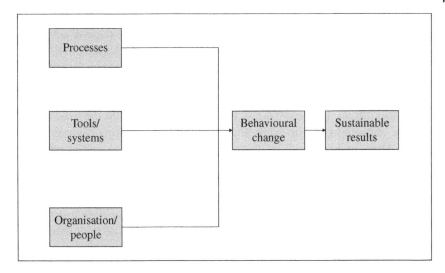

Figure 7.2 Processes, Tools/systems, and People leading to Sustainable results.

trust the outcomes. And if the workprocess around maintenance and updating the controller is not arranged, the controller will, after a certain time, work in a suboptimal way and may even lead to deterioration of the production output.

In the survey results of [7] we observed a broad range of applications in Process Industries. These industries tend to make a broad selection from the "bag of tricks" of control technology, possibly driven by their business needs for low-cost production and/or Time-to-Market. Overall, and not surprisingly, we see much focus on basic technologies; basic process control, PID, motion control, robotics, automated dosing systems. In comparison to the size of this industry sector, the use of more advanced control techniques is still quite limited. Traditionally, this was due to the lack of computational power, the high perceived costs, and the need to have qualified personnel on site to maintain the application. Most of these barriers, except the need for personnel, have been overcome in the past decades.

The surprising fact in the process industries is that most of the installations are equipped with control systems (DCS, Scada, PLC), but that the built-in capabilities of these systems that go far beyond basic control algorithms are hardly being used. A state-of-the-art Distributed Control System is capable of handling a unit-based Model Predictive Control (MPC) and advanced calculation models. Apparently, in this industry, the Tools/Systems installed are not the limiting factor for further exploitation of automatic control. This brings us to the insight that the hurdles for application are more in the Processes and People parts of Figure 7.2.

Characteristic for the process industry is that in general automatic control is accompanied by human intervention. In almost all plants, operators supervise the production processes, real autonomous operation (see Figure 7.9) is still very rare. In recent years many overview publications regarding the status of automatic control in the process industries have been published: examples are the NAMUR (German Process Industry Association) survey, studies reported by Samad et al., and an overview by Lamnabhi et al. [4, 10–12].

Very important for the actual application of automatic control in process industries is also the way it is organized as a discipline. Traditionally, the industry has in house engineering departments or makes use of engineering contractors for process design and construction. In these departments, process design generally starts with a process block diagram, where at the end of the design process some controls are "bolted on" at a later stage. It has been argued by many authors, e.g. in [13] and [14], that this is a suboptimal approach of designing a process with its controls, but it is still reality in most of the industries. In process industry operations, the control discipline is mostly organized on plant/site level with a big focus on troubleshooting and maintaining the assets. Innovations in automatic control come through suppliers or in some cases from internal corporate technology departments. The discipline of Process Systems Engineering (see, e.g. Sargent, Pistikopoulos et al., Klatt and Marquardt [15–17]) puts Automatic Control, traditionally a "hidden" technology, in the bigger picture of Systems thinking and systems design.

Landmark Innovations

Several important landmark innovations are recognized as originating in the process industry, such as the Steam governor, the application of MPC in (petro)-chemical installations [9], and Kalman filtering. For especially the bigger chemical plants, the application of a real-time optimizer, making use of rigorous process models to orchestrate the process units, constitutes a real landmark. In more recent times, we see the application of technologies for providing the process operators or maintenance technicians with up-to-date and reliable information (Human-Machine-Interface (HMI), tablets, and augmented reality).

Value Proposition

In terms of the perceived benefits of control technology, we questioned the industry on the average pay back time (see [7]). Thereby assuming that all positive effects of applying control technology, whether it is the reduction of energy consumption, the increased uptime, the decreased time-to market or the like, can be translated into money. Process-oriented clusters show the best pay back results. For a technology like MPC, 2–4% revenue increase was often used as a benchmark figure proven in practice, which for the petro- and bulkchemicals mostly led to an interesting return on investment. In specialty chemicals/Pharma, producing between narrow quality limits is essential, and leads to premium pricing and the avoidance of very costly call back/liability actions.

Digitalization

Throughout all process industries, a big theme during the last 5–10 years is Digitalization (sometimes also referred to as Industry 4.0 or Smart Manufacturing). First this was received in the industry with scepticism ("we are already applying digital technologies for 40 years"; see also the relatively low scores for digitalization as a driver in process industries: [7]). Later, in many companies, pilot projects were carried out, collecting all process data, production data, quantitative, or in text format. This often led to fruitful insights (e.g. regarding factors influencing product quality), but sometimes also to disappointing results. For example, in a chemical plant a "big data" project costing 1 million Euro resulted in the conclusion that a basic controller was not properly operating. An issue that also could (or should) have been solved with proper maintenance of the basic controls in that plant. Another flipside of the application of data analytics is the lack of interpretation in terms of physical/chemical relationships. This constitutes one of the most important scientific challenges for automatic control in process industries: Most often chemical/physical relationships are available, but process knowledge should be extended with data analytics and advanced sensing.

Digitalization also comprises the use of digital twins, advanced analytics and the Internet-of-Things, including advanced sensor technology [18]. Performance effects are a.o. summarised in [1]. That paper describes a conceptual framework for product and service innovation, thereby incorporating digital technology adoption, showing that companies with higher levels of digital technology implementation can introduce more radical product and service innovations.

Digitalization is often claimed to break down the traditional barriers in organizations. In process industries, automation was most often organized according to a hierarchical layered model (see Figure 7.3: The Automation Pyramid), with

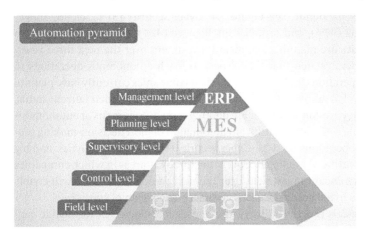

Figure 7.3 The automation pyramid.

basic control and safety systems as the lowest layer, unit control, plant control and production/business control and optimization as the upper layers. With digitalization, the availability and transparency of information throughout the company is almost unlimited: everybody may have access to accurate and timely data necessary to fulfill their work and/or control the production process. In that way, the traditional layers are broken down and the overall performance may increase. On the other hand, the basic safeguarding layer should be kept clean and separated (which is also required by the IEC61511 regulations for Process Safety).

Cybersecurity

Driven by incidents, ransomware attacks and the like, process industries have shifted focus toward improving the cyber resilience. The consequences for individual plants are often procedural (instructing people) but very often infrastructural: improving/replacing obsolete control systems which are more vulnerable and preventing the spread of impacts of incidents ("defense in depth": the zoning principle, conform the IEC 62443 standard series developed to secure industrial automation and control systems (IACS) throughout their life cycle [19].). The impact of cybersecurity is therefore somewhat dualistic: On the one hand an opportunity, by driving the replacement of obsolete systems. On the other hand, a threat, because the zoning principle hampers the breakdown of the traditional automation pyramid.

Future Investments

How does process industry currently look at investments in the area of automatic control?

A global survey conducted by Yokogawa [20] has revealed that process industry companies are accelerating investment in industrial autonomy in the background of the Covid-19 pandemic, see Figure 7.4. Cyber security (51%), cloud, analytics, and big data (47%), and artificial intelligence (42%) are three vital areas in which companies are planning significant investment over the next three years. 64% of respondents stated that they expect to reach autonomous operations in their primary operations by 2030. 89% said their companies currently have plans to increase the level of autonomy in their operations. Regarding their current status, 64% said that they are conducting or are piloting semi-autonomous or autonomous operations, while 67% expect significant automation of most decision-making processes in plant operations by 2023. Cyber security (51%), cloud, analytics, and big data (47%), and artificial intelligence (42%) are three key areas in which companies are planning significant investment over the next three years. These will enable organizations to make better decisions across a greater span of control.

For the chemicals process industry, the remarkable change in worldwide capital spending focus over the years was already referred to in Section 7.1.1. Since investments in automatic control are a significant part of capital spending, this will

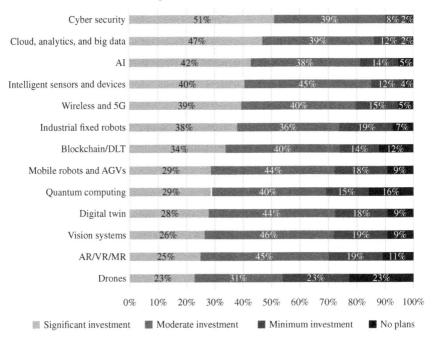

Figure 7.4 Level of investment in technologies for production processes over the next three years. Source: Yokogawa [20], published in Process Industry Worldwide/Yokogawa.

also lead to geographical consequences for the focus areas in automatic control for this industry.

Drivers for Automatic Control

Finally, the graphs of Figure 7.5 illustrate the relative ranking of drivers for automatic control in the energy, oil and gas industry versus the other subsectors

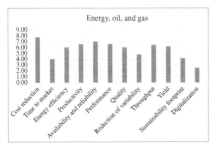
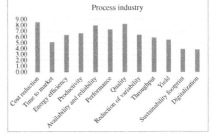

Figure 7.5 Ranking of main drivers for automatic control in Process Industry, Oil & Gas.

in process industry (as explained in [7]). It shows the importance of cost reduction, equipment availability, and reliability, as well as quality. Quite remarkably, digitalization does not stand out as a prime driver currently.

7.2 Existing Control Challenges in Process Industry

In this section, we will translate the current stand of automatic control in process industry into control-specific challenges. In Section 7.3, we will concentrate on vision driven innovation and the new challenges, and highlight the consequences for the control research problems in this sector.

7.2.1 Continuous/Batch

For continuous, single-grade processes, control design methods are well-established and developed. Special challenges occur in situations with multiple grades and/or varying feedstocks. Batch processing and especially the combination of continuous and batch steps, constitute extra challenges for optimal control design. Fermentative/biotech processes are a special category, being based on growth rates of living organisms. Finally, looking at mode switches between process operations, and changes in the process routing for different grades, we identify the challenge on how to adapt and design an overall best control scheme.

As mentioned in the introduction, we will use a continuous and batch example throughout this chapter. See Figures 7.6 and 7.7 for a schematic description. At the end of each section, the specifics for these examples will be discussed.

7.2.2 Broad Range in Dynamics

A special control challenge is the broad range in duration of the dynamic response of the process often seen. Many processes in this industry show slowly varying behavior, due to fouling, catalyst degradation and the like. On the other hand, chemical reactions may show runaway behavior under circumstances. Anti-surge control of compressors is another example where fast dynamic response is key.

7.2.3 Fault Detection and Abnormal Situation Management

This constitutes a very important focus area in process industry: what to do if a distillation column floods, a compressor surges, an instrument fails? Do we need to shutdown the plant or produce at lower capacity? And what are the consequences for the control strategy?

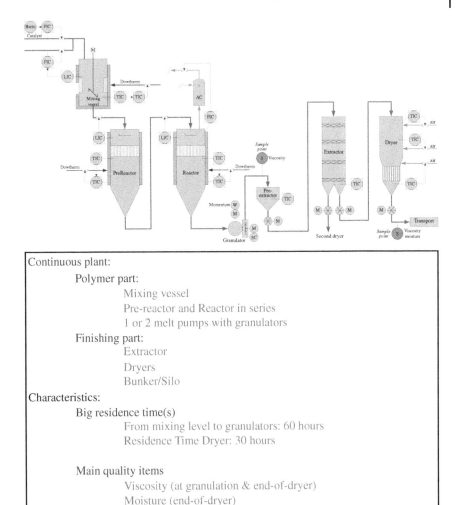

Continuous plant:
    Polymer part:
        Mixing vessel
        Pre-reactor and Reactor in series
        1 or 2 melt pumps with granulators
    Finishing part:
        Extractor
        Dryers
        Bunker/Silo
Characteristics:
    Big residence time(s)
        From mixing level to granulators: 60 hours
        Residence Time Dryer: 30 hours

    Main quality items
        Viscosity (at granulation & end-of-dryer)
        Moisture (end-of-dryer)
        Sampling: manually

Figure 7.6 Continuous plant example.

7.2.4 Grey Box Models

The availability of process knowledge regarding the unit to be controlled is a key success factor in process industry. A grey box model combines qualitative prior knowledge with quantitative data. This approach uses all available information about a certain industrial process to determine the best possible process model. Chemical/physical relationships and conservation laws provide a basis for a model, but most often, there are additional unknown parameters, such as

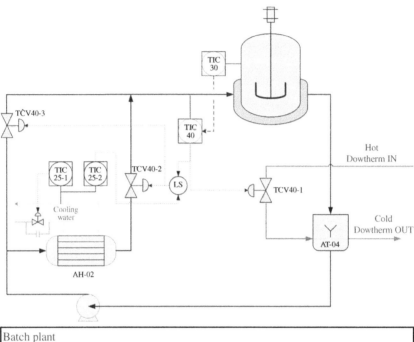

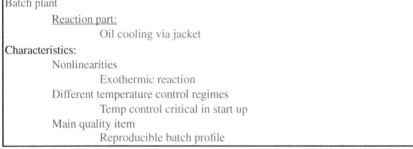

Figure 7.7 Batch plant example.

reaction kinetics, catalyst activity, growth rate of organisms, etc. These must be estimated, preferably online, to create a process model that can be applied to control and optimize the process. Interesting challenge for this industry is that there are a lot of nonlinear relationships, and hard constraints regarding, e.g. temperature, pressure, and flow, which may not be violated.

7.2.5 Multiple Objectives

In a production process the objectives may be to optimize throughput, yield, energy, ecofootprint or flexibility. Probably the most important challenge from a control perspective is multi-objective optimization: How do we create a control policy that maximizes throughput/yield while reducing energy consumption and, at the same time, minimize the emissions.

7.2.6 Control in the Design Process

As mentioned in Section 7.1.2, in most industries the control strategy is developed in the final stage of the process design. This results in bolt-on, rather than integrated control. And, as a consequence, often a suboptimal designed process, with over-dimensioning, intermediate stocks and the like. Integration of process design and process control is a key challenge and has been addressed in many papers already (cf [13, 16–17]).

7.2.7 Focus on the Basics

As mentioned before, many assets in process industry have a long lifetime, and the initially designed controls will be present for a long time. The challenge is here to keep this basic level of control performing adequately. In many on site studies it has been shown that up to 70% of the industrial controllers are not tuned or designed properly. Controller performance monitoring tools/methods are key.

7.2.8 Application to Continuous and Batch Examples

Here we apply the challenges of Sections 7.2.1 to 7.2.7 to the continuous and batch plant examples. Continuous plant example: This example exhibits many of the challenges listed above. The broad range in dynamics is obvious. The high residence times are combined with the need for good temperature and viscosity control. Multiple objectives for this plant may be relevant due to market conditions and energy/emission constraints. Batch plant example: Here the broad range in dynamics is obvious as well. Controlling the temperature at start up requires a scheduled tuning (which was initially not foreseen). Focus on the basics is required. And abnormal situation management, given the exothermic nature of the reaction, is key.

7.3 Vision of the Future Generation of Products and Processes

Referring to the innovation cycle described in Chapter 1 (see Figure 7.8), we identified a number of areas (process steps) to activate the cycle. First, the concept of vision-driven innovation (this section), and secondly, the systematic translation of industry challenges into control research problems/research directions, which will be the topic of Section 7.4.

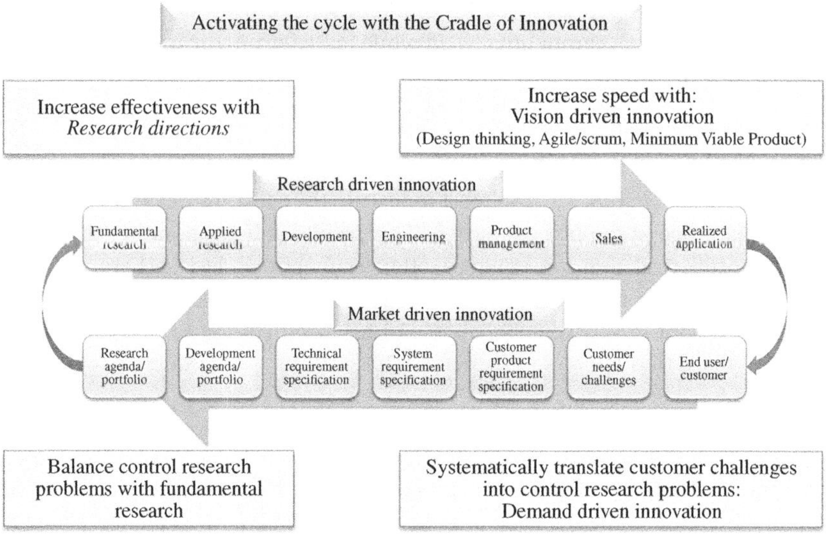

Activating the cycle with the Cradle of Innovation

Increase effectiveness with
Research directions

Increase speed with:
Vision driven innovation
(Design thinking, Agile/scrum, Minimum Viable Product)

Research driven innovation

| Fundamental research | Applied research | Development | Engineering | Product management | Sales | Realized application |

Market driven innovation

| Research agenda/ portfolio | Development agenda/ portfolio | Technical requirement specification | System requirement specification | Customer product requirement specification | Customer needs/ challenges | End user/ customer |

Balance control research
problems with fundamental
research

Systematically translate customer challenges
into control research problems:
Demand driven innovation

Figure 7.8 The innovation cycle.

7.3.1 Vision-Driven Innovation in Process Industry

The extent to which vision-driven innovation tools are applied is highly dependent on the subcategory of Process Industry. For example, in the Oil and Gas and chemicals industries, Design Thinking workshops or other approaches often result in an Operations Vision based on statements like:

- Autonomous operation
- Remote operation (e.g. for drilling platforms)
- Paperless plant
- Avoidance of manual labor (hazardous/unpleasant)
- Quality right first time
- Self-optimizing plant

Projects supporting such a vision are subsequently ranked according to cost/benefit and ease of implementation. In many cases, this is part of broader "Operational Excellence" or Lean programs. This is of course not directly focused on control solutions, and in practice, projects with an advanced automation component often die in an early stage due to risks, uncertainties, or lack of awareness. In some companies it has been made mandatory to carry out an automation assessment as part of Operational Excellence [21]. Some other companies even take this a step further: applying advanced control techniques is set as a standard (i.e. the plant needs to justify/explain why it is not applied). But these are still exceptions in process industry.

7.3.2 Process Industry Customer Challenges

Looking at the current application of automatic control in process industry and its existing challenges, the operations visions and new upcoming technologies, we can formulate the challenges of process industries, as a customer of control technology, into system/technical requirements and subsequently control research problems. As argued in Section 7.1, this is just one part of the story: People-Processes and Tools/Systems need to go hand-in-hand to achieve sustainable value generation. Our focus here is on the Tools/Systems aspects. However, it is important to stress the logical sequence that is found in any improvement project: i.e. define where you want to go and where you are, define how to reach this and how to keep this:

1. Start with a Vision; Make an automation plan based upon an assessment
2. Implement accordingly (see Figure 7.2: People-Processes-Tools/Systems)
3. Close the loop (i.e. maintain/improve what has been implemented, and learn from it for further technology innovation)

The customer challenges a to h are explained below and will be applied to the continuous and batch plant example at the end of this section. As referred to in Section 7.1, probably the biggest and overarching challenges for process industries for the next 10 years will be the Energy transition and the Reduction of the ecofootprint.

a. Energy transition: For the process industries this will most likely result in novel process designs with nonfossil fueled equipment. Control and automation aspects of the development of this equipment is not part of this chapter but part of Chapter 6: Robotics and Manufacturing Automation. For process industry the control consequences are limited to opportunities in the process design stage (see below c).
b. Reduction of the ecofootprint: For this industry the efforts will also result in novel process and plant designs, with increased complexity. New sensor technologies and modeling techniques are key supportive technologies.
c. Simultaneous process design and control: With a and b above, it is even more needed for automatic control to be in the front seat and seize the opportunity given. The fact that a major revamp of processes is ahead of us constitutes a huge opportunity for automatic control. Therefore, we will formulate this in Section 7.4 as a special research direction.
d. Data-driven engineering and vertical information integration: The German Process Automation End User Association NAMUR formulated additional

innovation challenges for the process industry as follows [4]: "Innovation is the practical implementation of ideas that result in the introduction of new goods or services or improvement in offering goods or services. There are a few fundamental drivers for innovation driven by an 'industry logic'. Innovation has to create an added value in the main business processes. These drivers are: Information integration along the Asset life cycle (design-operation-maintenance: Data-driven engineering); Vertical information integration from field level to business level; and Modularization/Process Orchestration for increased flexibility and agility." In data-driven engineering, the idea is to standardize and streamline the engineering process as much as possible with the aid of digital tools, with the aim of speeding up the design process and reducing costs. For vertical information integration, the idea is to aggregate and contextualize plant data in such a way that better decision-making at the higher levels can be achieved. Simulation and decision support techniques (What if…) play an important role.

e. Flexibility/agility through modularization/process orchestration: Modular automation with design of process units of high degree of intelligence and integrated control systems (black-box modules) with a super-ordinated orchestration is the present key enabler for flexible plant operation: Academia together with operating companies should develop, see [4]: – Fundamentals of decision-making in module-based planning and design. – Prototypes of simulation-based decision support tools that help to – Select modular equipment against the background of insecure data. – Take the decision for or against modular design in early design phases based on a sound economic comparison of conventional and modular design. – New, robust design approaches aiming at increasing equipment flexibility and process tolerance. Process Orchestration is the orchestration of intelligent process modules with manufacturer independent diagnosis.

Additional customer challenges from the different subcategories in Process Industry comprise:

f. Autonomous operations: As mentioned earlier, this is ranked by the majority of process industries as an area of investment for the next 10 years. At the highest level of autonomous operation, humans may not be needed for any intervention. Humans may be in remote locations, since their immediate presence is not required. An example of this would be a mining operation where the haul trucks operate without drivers. These vehicles would be able to detect when obstacles, such as humans, were in their path and avoid collisions on their way to deliver the ore. See Figure 7.9 for the levels definition of autonomous operations as applied also in SAE.

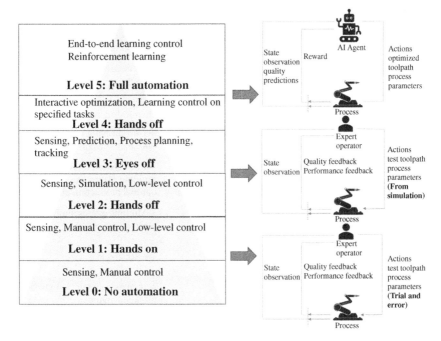

Figure 7.9 Levels in autonomous operations (see also SAE definitions in Chapter 4: Automotive Control and Chapter 6: Robotics and Manufacturing).

g. Miniaturization/process intensification: This focuses primarily on chemicals and the creation of optimal conditions for molecules to react. The direct benefits are a reduction of investments in equipment, a reduction of the energy consumption and emissions. Consequences for automatic control research needs are not directly foreseen.

h. Quality by design (QbD): This is a concept first outlined by quality expert Joseph M. Juran in publications, Designing for quality and innovation is one of the three universal processes of Juran, describing what is required to achieve breakthroughs in new products, services, and processes. While QbD principles have been used to advance product and process quality in industry, and particularly the automotive industry, they have also been adopted by the U.S. Food and Drug Administration (FDA, see [5]) for the discovery, development, and manufacture of drugs.

Application to Continuous and Batch examples: *Table 7.1 translates the customer challenges of this section to both the continuous and batch plant example.*

Table 7.1 Process Industry customer challenges and the two example processes.

Customer challenge	Continuous plant example	Batch plant example
Energy transition	++ Energy intensive process with lots of heat integration	+ Moderate energy use
Reduction of eco footprint	+/− Moderately relevant	+ Relevant given the chemicals used
Simultaneous process design and control	+ Current plant designed with bolt - on controls	+ Current plant designed with bolt - on controls
Data driven engineering and vertical information integration	+ Quite relevant given market demands on quality information and certification	++ See vision statements, and multiple grades requiring optimal production planning
Flexibility/Agility through modularization/ orchestration	− Less relevant, single product plant	+ Multiple grades, multiple reactors in parallel
Autonomous operation	++ See above	++ See above
Miniaturization/Process Intensification	−− Less relevant	+ Increasingly relevant given market demands
Quality by design	+ Increasingly relevant for this market (customers mainly automotive/ electronics industry)	+/− Moderately relevant

Continuous plant example: For this plant, a vision-driven innovation workshop was carried out. Main results:

- *Autonomous operation*
- *Avoidance of manual labor (hazardous/unpleasant)*
- *Self-optimizing plant*

Batch plant example: For this plant also a vision-driven innovation workshop was carried out. Main results:

- *Paperless plant*
- *Avoidance of manual labor (dosing)*
- *Quality right first time*

7.4 Formulation of Control Research – Directions for Process Industry

In this section, we will translate the vision statements and customer challenges from Section 7.3 into control requirements and control research directions. Please note that some vision statements and challenges are being supported by more than one research direction.

7.4.1 Research Directions

We will make a distinction between Strategic tools, tools for Operations and tools for Design/Engineering. For the strategic part, we refer to the logical sequence explained in Section 7.3.

Start with a Vision; Make an automation plan, based upon an assessment; Implement accordingly and close the loop.

This leads to the following research directions d1–d3:

d1. Vision-driven innovation tool development: Motivation: In this, by nature, conservative industry sector, it is of prime importance to support the local vision development process with state-of-the-art automatic control tools/prototypes. The objective would be to develop a toolkit with which a vision development (or design thinking) session on a production site can be carried out, in conjunction with similar exercises on digitalization and operational excellence (which in a lot of companies are already being carried out). The expected breakthroughs are a correct positioning of automatic control solutions next to other routes to operations improvement, and as such a higher valuation and degree of implementation of the technology.

d2. Maturity assessment and benchmark development: Motivation: As referred to in Section 7.3, in several companies an automation assessment is part of the general improvement/innovation process [21]. In such an assessment the plant under study is scored compared to international benchmarks, on the degree of automation, control performance, alarms in operations, and other measurable quantities, including aspects of engineering, maintenance and organization. Achieving a certain score on the benchmark scale is being made part of the Operational Excellence targets. Sharpening such an assessment with the latest technological insights is definitely a key research topic. The objective is to develop a maturity assessment approach for automatic control. With a crisp definition of the underlying metrics. Part of the assessment should also be a list of recommendations/next steps (which technologies and best practices to apply when) to improve the assessment score. The expected breakthroughs are a clear focus and awareness in process industry on the competitive position and as such also a big improvement in selecting the right degree of automatic control.

d3. The automation plan: Motivation: As mentioned before, the lack of a dedicated and thorough multiyear plan for automatic control projects is one of the main causes of the lack of focus. The gaps coming from the vision and the maturity assessment will result in a list of potential projects to execute. The challenge is to select and place these projects in a logical sequence. And to be able to do a reasonable estimate of the costs/benefits and the required resources. To support this direction, data driven engineering (described in d7) is key. The objective is to provide process industry with automation project selection and planning tools, to arrive at a correct and adequate implementation. The expected breakthrough is that automatic control is really on the agenda of operations management and, as a result, will be applied in more cases.

The following problems/research directions are related to Process Industry Operations, the Value chain:

d4. White models and data analytics: digital twins: Motivation: To create models that can be applied for direct plant control and control design, it is key to make use of the existing plant knowledge and fundamental chemical/physical relationships. Interesting challenge is for this industry that there are a lot of nonlinear relationships, and hard constraints regarding, e.g. temperature, pressure, and flow, which may not be violated. On top of this, there are additional unknown parameters, like reaction kinetics, catalyst activity, growth rate of organisms, etc. These have to be estimated, preferably online, to create a process model that can be applied to control and optimize the process. Data analytics tools and techniques provide additional ways of estimating these parameters and potentially generating additional insights. The objective is to enhance current modeling tools with big data analytics. The expected breakthroughs are the creation of real digital twins that can be used for control design and optimization, that may also serve other purposes.

d5. Process orchestration: hierarchical optimization: Motivation: In most process industries flexibility and agility are required. To that end, process equipment, modules, and routing may be changed from one production run to the other. The challenge is to be able to produce in an optimal way under these varying circumstances. This requires optimization techniques able to handle the orchestration of process units. Reference can be made to the European Union project CoPro [22]. The objective is to develop robust optimization techniques that can generate optimal settings for equipment control. The expected breakthroughs are more production flexibility, less downtime, and less intermediate storage.

d6. Dynamic optimization for design and for autonomous operation Motivation: To be able to handle the increasing amount of nonstationary processes in this industry, dynamic optimization is the best solution. Traditional drawbacks of computing power and dimensionality have disappeared to a large extent, and now applications of dynamic optimisation are within reach. These applications may exist in optimal plant design, for autonomous operation (the self-optimizing plant) as well as in the optimal trajectory of batch processes. The challenge will be to derive robust and maintainable control strategies for use in industrial environments. The objective would be to develop robust dynamic optimization techniques for use in collaboration with digital twins for plant design, and for control strategy design in operations. The expected breakthroughs are an increased quality of plant design and sizing, as well as online optimization of throughput/yield/energy in operations.

For Process Design/Engineering, or Asset lifecycle, the following research directions apply: Digital twins: see above research direction d4.

d7. Data-driven engineering to increase speed and flexibility: Motivation: In process industry, the engineering of complex installations is still a cumbersome process, similar to a craft. With the aid of engineering tools, digital twins and templates for control design, an increase in speed and effectiveness is achievable. Using a coherent set of engineering data definitions, modular designs, templates, etc., will also pay off in the operations and maintain phase of an installation. Work on this topic has already started in some subcategories of process industry in Europe but should be expanded. The objective is to develop a standard data-driven engineering environment for modular plant design. The expected breakthroughs are more speed in plant design and upgrading, as well as more consistent application of best practices, which is required with the challenges of energy transition and ecofootprint reduction.

d8. Interaction and integration of process operation, design, and control: Motivation: This topic is sometimes also referred to as Process Systems Engineering, or Systems thinking: the close cooperation between Process Technology and Process Control, cf [13, 17]. In industrial engineering practice, overall plant design is still being carried out in three independent sequential stages; (i) choice of operating mode, (ii) process design and (iii) process control. The (future) challenges for process control include: process innovation including intensification (is the new process controllable?), dynamic operation (e.g. reverse flow reactors), flexible operation (with regard to products and production volumes), higher performance, tighter constraints, and more severe disturbances (e.g. feedstock variation). Therefore, controlling the plant dynamics is becoming of vital importance. These challenges require interaction and integration of process design and control, as the dynamics of a controlled plant are not only determined by control but also by the uncontrolled plant dynamics. The objective is to develop an approach, theory, tools, etc., to allow for better interaction and even integration of process operation, design and control. In the case of better interaction, one can think of making sure that the relevant aspects of stage (iii) are taken into account during stage (ii). Integration means that at least two stages are done at the same time. So, control is to be considered as an integral part of the process and equipment design. The expected breakthroughs are: Shorter process design times. Higher material and energy efficiency.

7.4.2 Consequences for Education in Automatic Control for Process Industry

From the above research directions, it can be concluded that it is paramount for students/practitioners to understand the role of automatic control in the bigger picture of process industry design and operations. The discipline of Process Systems Engineering includes control in the overall design of production processes (ref. [15, 23]). Pistikopoulos [16] treats the multi-layered view of process systems engineering. Klatt and Marquardt [17] treat academic and industrial perspectives on the research and applications of PSE, focusing systems thinking and systems

Table 7.2 Control problems/research directions applied to the two example processes.

Control problem/ research direction	Continuous plant example	Batch plant example
Vision driven innovation tool development	+ Relevant given rapid developments in automation tools and technologies	+ Relevant given rapid developments in automation tools and technologies
Maturity assessment and benchmark development	++ Important because few similar plants exist	+ Many similar installations for this particular process technology
The automation plan	++ Many activities concentrated in plant shutdown	+ Lots of activities on-the fly
Combination of white models and enhancement with big data analytics: digital twins.	++ Extension of current knowledge of reaction kinetics: high benefits perceived	+ Development/improvement of golden batch profile
Process Orchestration: Hierarchical Optimization	+/− Moderate value due to single grade process	+ High value perceived due to multiple grades/streams
Dynamic Optimization for design and autonomous operation	++ Given complexity of current process for operators	+ Relevant for autonomous operations given current complicated temp. control
Data driven engineering to increase speed and flexibility	+/− Moderately relevant	+ High value returns expected
Interaction and integration of process operation, design and control	+ Savings in equipment sizing as well as reduction of energy/ecofootprint	++ Important to completely rethink the design of the plant for the future

problem solving. Udugama [18] treats the consequences of digitalization for educating engineers in the field of process and chemical engineering.

Application to Continuous and Batch examples: *Table 7.2 translates the Control Problems and Research directions formulated in this section to both the continuous and batch example.*

7.5 Conclusion: What Drives and Blocks Innovation in Process Industry?

Activating the flow in the innovation cycle (see Figure 7.8) is key for automatic control as a pivotal discipline to address societal and technological challenges of our future. This requires not only novel tools, theories, and trained people but also innovative ways of working together.

For Process Industry, it can be concluded that especially vision-driven innovation is to be fostered: there is obviously a need for speed, given the imminent challenges of the energy transition and the reduction of the ecofootprint. Cooperation mechanisms through the enhanced collaboration of suppliers-institutes-user associations will help the systematic translation of customer challenges into control research problems. This is also highlighted in an IFAC publication of Serbezov et al. [24].

The research problems/directions listed in Section 7.4 form a starting point to focus the research agenda as far as process industry is concerned. This is, however, a dynamic list: it will change with new developments and changes in the process industry ecosystem. The question is what specific *fundamental* research would be required to support these directions. This is something for academia to address in consultation with the industry. Consequences for the curriculum should also be evaluated.

The question is how this applies to the two example processes used in this chapter. *By providing the relevance for both plant types (see Table 7.2), it can be concluded that the drivers and blockers for innovation are of course in detail dependent on the plant in question, but that many similarities exist. This gives a good foundation for the framework applied and the resulting research directions.*

The biggest drivers for innovation regarding Automatic Control in Process Industry are:

- The need to fundamentally rethink process operations due to the energy transition and ecofootprint reduction.
- Digitalization and the associated need for more agility and flexibility.
- The value potential by increasing the level of automatic control.

The biggest hurdles for innovation are:

- Focus on the existing assets, and a lack of awareness and recognition of what control can bring.
- Control is introduced in a too-late stage in the design process.

In terms of People-Processes-Tools/Systems, the following recommendations apply:

People:

- There is a need for more cooperation and knowledge exchange to improve awareness both in academia and industry.
- Sustainably implementing automatic control requires the people aspect to be taken into account in the development and design stage.

Processes:

- Start with a Vision; Make an Automation Plan based upon an Assessment.
- Formation of consortia with academia, system integrators, technology suppliers, and end-user companies. Organization like IFAC can actively coordinate/connect, cf [24]. The list of research directions in this chapter can be used as a starting point and structurally updated.
- Share success stories of cooperation/ecosystems.

Tools/Systems:

- The pertaining research directions of Section 7.4 are to be actively managed in a programmatic approach.
- In general, when developing tools/systems, focus more on the implementation aspects and link them with Digitalization/Operational Excellence approaches.

Process Industry innovation is highly impactfull toward a sustainable future. Balancing the right research directions in Tools/Systems, Processes, and People will support Automatic Control innovation to play the role needed to face societal and technological challenges.

References

1 Henrik Blichfeldt and Rita Faullant. Performance effects of digital technology adoption and product and service innovation –a process-industry perspective. *Technovation*, 105:102275, 2021. doi: 10.1016/j.technovation.2021.102275.

2 Thomas Lager. Managing innovation & technology in the process industries: Current practices and future perspectives. *Procedia Engineering*, 138:459–471, 2016. doi: 10.1016/j.proeng.2016.02.105.

3 The European Chemical Industry Council (CEFIC). The European Chemical Industry: A Vital Part of Europe's Future. Facts & Figures, 2021. https://cefic .org/a-pillar-of-the-european-economy/.

4 W. Otten and F. Hanisch. Automation –the driver of smart production. In *Smart Process Manufacturing Kongress*, pages 1–30, Germany. Oct 27–28, 2020.

5 U.S. Food and Drug Administration. Food and drug administration guidelines, 2023. URL https://www.fda.gov/.

6 Thomas Lager. A conceptual analysis of conditions for innovation in the process industries and a guiding framework for industry collaboration and further research. *International Journal of Technological Learning, Innovation and Development*, 9:189, 2017. doi: 10.1504/IJTLID.2017.10008245.

7 Silvia Mastellone and Alex van Delft. The impact of control research on industrial innovation: What would it take to make it happen? *Control Engineering Practice*, 111:104737, 2021.

8 James Blake Rawlings and David Q. Mayne. *Model predictive control: Theory and design*. Nob Hill Pub., 2009.

9 Michael Baldea, Juan Du, Jungup Park, and Iiro Harjunkoski. Integrated production scheduling and model predictive control of continuous processes. *AIChE Journal*, 61(12):4179–4190, 2015.

10 V. Hagenmeyer and U. Piechottka. Innovative process operation and control-experiences and perspectives. *Automatisierungstechnische Praxis*, 1–2:48–64, 2009.

11 Tariq Samad, Margret Bauer, Scott Bortoff, Stefano Di Cairano, Lorenzo Fagiano, Peter Fogh Odgaard, R. Russell Rhinehart, Ricardo Sánchez-Peña, Atanas Serbezov, Finn Ankersen et al. Industry engagement with control research: Perspective and messages. *Annual Reviews in Control*, 49:1–14, 2020.

12 Francoise Lamnabhi-Lagarrigue, Anuradha Annaswamy, Sebastian Engell, Alf Isaksson, Pramod Khargonekar, Richard M. Murray, Henk Nijmeijer, Tariq Samad, Dawn Tilbury, and Paul Van den Hof. Systems & control for the future of humanity, research agenda: Current and future roles, impact and grand challenges. *Annual Reviews in Control*, 43:1–64, 2017.

13 J.E. Rijnsdorp. *Integrated process control and automation*. Elsevier Science Publishers, 1991.

14 A.G.E.P. van Delft. *Dynamic optimisation of thermal energy systems*. PhD thesis, Eindhoven University of Technology, 1989.

15 R.W.H. Sargent. Advances in modelling and analysis of chemical process systems. *Computers & Chemical Engineering*, 7(4):219–237, 1983. doi: 10.1016/0098-1354(83)80013-1.

16 E.N. Pistikopoulos, Ana Barbosa-Povoa, Jay H. Lee, Ruth Misener, Alexander Mitsos, G.V. Reklaitis, V. Venkatasubramanian, Fengqi You, and Rafiqul Gani.

Process systems engineering–the generation next? *Computers & Chemical Engineering*, 147:107252, 2021. doi: 10.1016/j.compchemeng.2021.107252.

17 Karsten-Ulrich Klatt and Wolfgang Marquardt. Perspectives for process systems engineering—personal views from academia and industry. *Computers & Chemical Engineering*, 33(3):536–550, 2009. doi: 10.1016/j.compchemeng.2008.09.002.

18 Isuru A. Udugama, Christoph Bayer, Saeid Baroutian, Krist V. Gernaey, Wei Yu, and Brent R. Young. Digitalisation in chemical engineering: Industrial needs, academic best practice, and curriculum limitations. *Education for Chemical Engineers*, 39:94–107, 2022. doi: 10.1016/j.ece. 2022.03.003.

19 International Electrotechnical Commission. IEC62443 Standard Series for Cybersecurity of Industrial Automation and Control Systems (IACS) Throughout their Lifecycle.

20 Ahlam Rais (ed). Two-thirds of process industry companies anticipating full autonomous operations by 2030: Yokogawa survey report. *Process Industry Worldwide*, 9, 2020.

21 WIB. Process automation maturity model. *WIB Process Automation End User Association report*, 2813, 2020.

22 Project team EU project CoPro. EU horizon 2020 project CoPro: Improved energy and resource efficiency by better coordination of production in the process industries: Final report, 2019.

23 Daniel R. Lewin, Emilia M. Kondili, Ian T. Cameron, Grégoire Léonard, Seyed Soheil Mansouri, Fernando G. Martins, Luis Ricardez-Sandoval, Hirokazu Sugiyama, and Edwin Zondervan. Agile process systems engineering education: What to teach, and how to teach. *Computers & Chemical Engineering*, 170:108–134, 2023. ISSN 0098-1354. doi: 10.1016/j.compchemeng.2023.108134.

24 A. Serbezov, R.R. Rhinehart, P. Goupil, and D.A. Anisi. Academic-practice collaborations in automation and control: Keys for success. *IFAC-PapersOnLine*, 55:308–313, 2022. doi: 10.1016/j.ifacol.2022.09.297.

Index

Legend: **boldface** refers to the most important explanation, <u>underlined</u> refers to figures in the text.

a

adaptive control   25, <u>29</u>, 78, 94, <u>105</u>, 163
additive manufacturing   172
advanced automation/control   160, 169, 212
agile approach   9
alarm(s)   52, 217
anomaly. *see* fault
artificial intelligence (AI)   15, 78, 95, 175, 201
  model   19
augmented reality   <u>95</u>, 189, 204
automatic control   1
automation assessment   94, 172, 186, 212
automotive control   3, **85**, 169
autonomous
  agents   182
  cars   4, 170
  manufacturing   171
  microgrids   4
  operation   206, 212, **215**
autonomy   104, <u>105</u>

b

batch process   171, 199, 208
benchmark(s)   204, 217
black-box model   177
building automation   3, 23, **43**

c

calibration   72, 93, 110, 172
carbon
  footprint   21, 29
  neutral   63
ChatGPT   17, 20
climate change   33, 61
co-design   22, 28, **32**
collaboration academia and industry   80, 221
compliance   94, 201
compressors   **46**, 53, 208
connected automated vehicles   112
continuous process   180, 199, **209**
control
  adaptive (*see* adaptive control)
  distributed (*see* distributed control)

The Impact of Automatic Control Research on Industrial Innovation: Enabling a Sustainable Future, First Edition. Edited by Silvia Mastellone and Alex van Delft.
© 2024 The Institute of Electrical and Electronics Engineers, Inc. Published 2024 by John Wiley & Sons, Inc.

control (*contd.*)
  feedback (*see* feedback control)
  hierarchical (*see* hierarchical control)
  model predictive (*see* model predictive
    control)
  nonlinear (*see* nonlinear control)
  optimal (*see* optimal control)
  real-time (*see* real-time control)
  state-space (*see* state-space control)
  supervisory (*see* supervisory control)
control and design integration/
    optimization   211, 213, 219
control design   10, 28, 54, 58, 62, 107,
    141, 162, 182, 208
control engineering   4, 10, 17, 36
critical infrastructure   5
curriculum   10, 221
customer requirements   6
cyber-physical systems   5, 15, 85, 169,
    179
cybersecurity   36, 78, 176, 206

d
data
  analytics   54, 172, 175, 187, 201, 218
  centers   **18**–34
  industry   3, 15
data-driven
  control design   163, 170, 178
  decision making   36
  engineering   213, 218
  models   173
decarbonization   62, 88
demand-driven innovation   8
design thinking   6, **7**, 212
diagnostics   51, 71, 87, <u>95</u>, 99, 104, 162
digitalization   8, 169, 182, 205
digital twin   **70**, 75, 94, 162, 172, 175,
    185, 205, 218
digitization   15
discrete   55, 141, 162, 200

disruptive innovation   7
distributed
  control   5, 35, 182
  control system (DCS)   203
  learning   94
dynamic
  coupling   66, 141
  model   49, 54
  optimal control   28
  optimization   218

e
eco footprint   20, 200, 213
edge computing   17, 36, <u>95</u>, 175
education   10, 11, 16, 220
electric vehicle(s)   35, 85, 97, 103
electrification   62, 86, **96**, 139, 148
emissions   43, 65, 85, **87**, 200, 211
enabling technology   37, 90, 182
energy   5, 9
  consumption   18, 43, 200, 210
  efficiency   60, 139, 145, 169
  footprint   18
  management   85, 99
  reduction   113
  transition   11, 200, 213
engineering
  data-driven (*see* data-driven
    engineering)
  systems (*see* systems engineering)
environmental   2, 87, 115
estimation
  model based   92
  of state (*see* state estimation)
ethics   16, 37
extended Kalman filter   71, 100, 184,
    204

f
fault (or anomaly)
  detection   54, 172, 208

tolerant control   109, 149
feedback control   26, 48, 87
first principles-based model
      (white model)   144, 162, 177,
      218
fuel cells   146
fundamental research   4, 6, 138, 164,
      212

g
gain scheduling   66
grey model   209
grid
   connected   144
   interactive   76

h
healthcare   5
heating, ventilation, and air-conditioning
      (HVAC)   **43**, 159
hierarchical control/optimization   25,
      50, 177, 186, 218, 220
human-
   -in-the-loop   176
   -machine interaction   175, 204
   -robot interaction   176
   -systems   5
human intervention   203
human labor   18
hybrid architecture   90
hybrid model   57, 79
hydrogen (H$_2$)   4, 91, 101, 103

i
identification. *see* system identification
incremental innovation   7, 170
industry
   data (*see* data industry)
   discrete (*see* discrete)
   process (*see* process industry)
Industry 4.0   4, 15, 182, 205

information and communication
      technology (ICT)   5, 112
information integration   109
innovation
   demand-driven (*see* demand-driven
      innovation)
   disruptive (*see* disruptive innovation)
   incremental (*see* incremental
      innovation)
   market-driven (*see* market-driven
      innovation)
   research-driven (*see* research-driven
      innovation)
   technological (*see* technological
      innovation)
   vision-driven (*see* vision-driven
      innovation)
integrating control and design. *see*
      control and design integration

l
layer of control/optimization   28, 50,
      106, 142, 159, 205
learning-based control   178
legacy equipment/systems   9, 76, 149

m
machine learning   25, 94, 109, 170
maintenance   20, 43, 52, 95, 163, 201,
      214
manual control   104, 171
manual labor   43, 72, 212
manufacturing   5, 18, 20, **169**
   automation   3, **169**
market-driven innovation   5
maturity   9, 89, 108, 138
   assessment   217, 220
microgrids   4, 22, 75, 147, 154
mobility   3, 17, 115, 138
model
   AI (*see* AI model)

model (*contd.*)
  black-box (*see* black-box model)
  dynamic (*see* dynamic model)
  first principles-based (*see* first
    principles-based model)
  grey (*see* grey model)
  physics-based (*see* physics-based
    model)
model-based control 44, 93, 163
model predictive control (MPC) 29, 69,
    71, 74, 90, 108, 174, 202
modular, modularity 29, 139
  design 32
  software architecture 107
motion control 183, 203
multi-objective control/optimization
    35, 115, 161, 211

n
neuromorphic 26
nonlinear, nonlinearity 33, 55, 66, 100,
    141, 183, 210
  control 162

o
operator (grid, plant) 9, 20, 30, 54, 65,
    159, 171, 202
  information 175
optimal control 25, 90, 93
optimization (of models/parameters)
  dynamic (*see* dynamic optimization)
  hierarchical (*see* hierarchical
    optimization)
  multi-objective (*see* multi-objective
    optimization)
  trajectory (*see* trajectory
    optimization)
orchestration 25, 137, 142, 201, 204,
    214, 218
  scenario-based 31, 35, 75

p
Photovoltaic(s) (PV) 29, 30, 44, 59, 75,
    138, 146
physics-based model 62, 79, 99, 180
PID control 4, 50, 108, 160, 182,
    203
power
  conversion/converter 3, 9, **137**, 139,
    141
  grid 54, 101, 137–155
  plant 180
  usage efficiency (PUE) 21
predictive maintenance 163, 172
process
  automation 15
  control/optimization 50, 154, 174,
    177, 203, 211, 219
  industry 3, **199**
  intensification 215, 216
  systems engineering 204, 219
programmable logic controller (PLC)
    53, 175, 203

q
quality 9, 28, 93, 145, 155, 171, 182,
    201, 208, 212
quality by design (QbD) 215

r
real-time control/optimization 59, 69,
    96, 170, 177, 182, 204
receding horizon control 28, 35, 174,
    185
reinforcement learning 25, 78, 110,
    170, 171, 215
reliability 7, 9, 26, 29, 71, 86, 144, 164,
    202, 207
renewable energy 65
requirements
  customer (*see* customer requirements)

safety (*see* safety requirements)
system (*see* system requirements)
technical (*see* technical requirements)
research
  directions   3, 9, 145, 162, 178, 199,
    216
  fundamental (*see* fundamental
    research)
  roadmap   5, 162
research-driven innovation   5
resilience   21
robotics   3, 9, **169**, 203
robust
  control   1, 27, 69, 108, 178, 218
  stability   51
robustness   53

s

safety   62, 94, 109, 144, 173, 201, 205
  requirements   9
  standards   110
scheduling   28
security   9, 21, 43, 146, 188
self-learning   169, 179
self-optimizing   170, 212, 218
sensing/sensor(s)   26, 32, 52, 65, 85, 88,
    93, 95, 172, 177, 181, 207
  soft (*see* soft sensing)
  virtual (*see* virtual sensing)
smart
  manufacturing   205
  mobility   139
societal challenges   5
soft sensing   180
stability   53, 139, 169
state estimation   57, 70, 73, 143, 162,
    185
state-space control   108, 141
supervisory control   44, 53, 113, 114,
    206

sustainable/sustainability   1, 20, 162,
    169, 187, 222
  development goals   11, 200
  footprint   9
  future   **1**, 164
system(s)
  architecture   47, 175
  dynamics   53
  engineering   5, 26, 37, 204, 219
  identification   162
  requirements   5
  thinking   22, 204, 219

t

technical requirement(s)   1–7, 10, 147,
    159, 213
technology
  enabling (*see* enabling technology)
  roadmap   170
  transfer   10, 110
technology/technological innovation
    1–2, 5, 170
test/testing   59, 87, 94, 110, 148, 172
training   9, 19, 25, 110
trajectory optimization   182–186, 218
tuning   74, 93, 154, 175, 185, 211

u

uncertainty   17, 69, 77, 110, 172

v

validation   110, 174
vehicle control   106
verification   174
virtual sensing   71, 93
vision-driven innovation   6, 79

w

water   5, 11, 47
wind energy   65, 143–146

Books in the IEEE Press Series on Control Systems Theory and Applications

Series Editor: Maria Domenica Di Benedetto, University of l'Aquila, Italy

The series publishes monographs, edited volumes, and textbooks which are geared for control scientists and engineers, as well as those working in various areas of applied mathematics such as optimization, game theory, and operations.

1. *Autonomous Road Vehicle Path Planning and Tracking Control*
   Levent Güvenç, Bilin Aksun-Güvenç, Sheng Zhu, and Sükrü Yaren Gelbal

2. *Embedded Control for Mobile Robotic Applications*
   Leena Vachhani, Pranjal Vyas, and Arunkumar G. K.

3. *Merging Optimization and Control in Power Systems: Physical and Cyber Restrictions in Distributed Frequency Control and Beyond*
   Feng Liu, Zhaojian Wang, Changhong Zhao, and Peng Yang

4. *Dynamic System Modeling and Analysis with MATLAB and Python: For Control Engineers*
   Jongrae Kim

5. *Model-Based Reinforcement Learning: From Data to Continuous Actions with a Python-based Toolbox*
   Milad Farsi and Jun Liu

6. *Disturbance Observer for Advanced Motion Control with MATLAB/Simulink*
   Akira Shimada

7. *Control over Communication Networks: Modeling, Analysis, and Design of Networked Control Systems and Multi-Agent Systems over Imperfect Communication Channels*
   Jianying Zheng, Liang Xu, Qinglei Hu, and Lihua Xie

8. *Advanced Control of Power Converters: Techniques and Matlab/Simulink Implementation*
   Hasan Komurcugil, Sertac Bayhan, Ramon Guzman, Mariusz Malinowski, and Haitham Abu-Rub

9. *Sensorless Control of Permanent Magnet Synchronous Machine Drives*
   Zi Qiang Zhu and Xi Meng Wu

10. *The Impact of Automatic Control Research on Industrial Innovation: Enabling a Sustainable Future*
    Silvia Mastellone and Alex van Delft

Printed and bound by CPI Group (UK) Ltd, Croydon, CR0 4YY

16/04/2025

14658345-0001